For Sotiria Tzortzaki, Pavlos Gletzakos, Foteini Bechraki and Panagiotis Gletzakos, teetering companionably with me on the edge of Europe, and great experts in the art of being alive

THE EDGES OF THE WORLD

www.penguin.co.uk

Also by Charles Foster

Being a Beast
Being a Human
The Screaming Sky
Cry of the Wild

THE EDGES OF THE WORLD

At the Margins of Life, Lands and History

Charles Foster

doubleday

TRANSWORLD PUBLISHERS

UK | USA | Canada | Ireland | Australia
India | New Zealand | South Africa

Transworld is part of the Penguin Random House group of companies
whose addresses can be found at global.penguinrandomhouse.com.

Penguin Random House UK, One Embassy Gardens, 8 Viaduct Gardens,
London SW11 7BW

penguin.co.uk

Penguin
Random House
UK

First published in Great Britain in 2026 by Doubleday
an imprint of Transworld Publishers

001

Typeset in 11/15.25 pt Sabon Next LT Pro by Six Red Marbles UK, Thetford, Norfolk
Printed and bound in Great Britain by Clays Ltd, Elcograf S.p.A.

The authorized representative in the EEA is Penguin Random House Ireland,
Morrison Chambers, 32 Nassau Street, Dublin D02 YH68.

A CIP catalogue record for this book is available from the British Library.

ISBN: 9780857529398

All my life I have loved edges; and the boundary-line that brings one thing sharply against another. All my life I have loved frames and limits; and I will maintain that the largest wilderness looks larger seen through a window.

G. K. CHESTERTON
Autobiography[1]

The world is a very narrow bridge, and the main thing is not to make yourself afraid.

REBBE NACHMAN OF BRESLOV

The quest of the Inner Ring will break your hearts unless you break it. But if you break it, a surprising result will follow.

C. S. LEWIS,
The Inner Ring[2]

Contents

Contents

A Word Before We Set Off

O NCE, IN A CHURCH, someone seemed to be saying that it was good to be wretched: to be on the edge of life, comfort and financial viability. It was the opposite of everything my education and my sybaritic instincts had taught me.

I wondered about this. As I did, I noticed that I was attracted to geographical, sociological and intellectual edges. This wasn't – or wasn't just – a reaction against my own centrism, and wasn't – or wasn't just – a form of Orientalism. It wasn't a morbid fascination with the exotic. As I looked (the look took years), it increasingly seemed that there was a general principle in play: a really, really general principle, spanning ethics, biology, politics, philosophy, cookery and just about every other domain.

The principle began with two observations: first, that edges were where everything significant happened; and second, that extreme goodness was to be found at the extreme edges, and the goodness was increasingly diluted as you retreated back towards the centre. I knew that correlation was not causation, but tentatively concluded that perhaps edginess *was* goodness, and goodness *was* edginess. Then I wandered away from the idea for decades and let it stew.

Over the years, I returned to it, and worried away at it for a bit, and then got distracted. I wrote bits and pieces in notebooks. I read commentaries on the Sermon on the Mount. I did many journeys in places I didn't want to be. I watched my children grow and asked, like every parent, 'How did this happen?' and 'Who are these people?' I became part of the ecosystem in a taverna where old Greek sailors went to die. I knew I had to honour them by writing something, but other things got in the way. And then, in Kenya, a long way from cosy

Oxford, watching destitute people walking at dawn along the edge of the main road into Nairobi, laughing, smiling and singing hymns, I felt compelled to pick the idea up again and take a proper look.

This book is the result. It tries not to degenerate into a rant against neoliberalism or homogenization – though I can't disguise my loathing of the metastasis of bloody mobile phone shops and corporate sandwich joints into every street in the world. It's not a systematic treatise. It is not a systematic anything. Still less is it a history of ideas or events viewed through the prism of edginess. Both those books could be written, but they should not be written by me. It is several other things. No doubt too many other things. It would have been easier, and very possibly prudent, simply to describe the view from various edges, compare those views with the (non) view from the centre, and draw some general moral and political conclusions from the comparison. But that didn't begin to do justice to what I was seeing.

The book is a patchwork of splashed impressions; a series of approximate cross-bearings which, taken together, identify the rough location of an idea. For every one of my examples, you can no doubt think of a counter-example. Good! For that shows the messiness of things: the inadequacy of received categories; the redundancy of top-down planning; the failure of edicts issued by central office to deal with the way things really are. Please humour me for a while, and stand back. There are plenty of dissonances, but you might find that they resolve into a chord that you recognize from a long time ago. They do for me.

No doubt I often overstate. I don't worry too much about that, for two reasons. First, if there's anything to this edge thesis, it's less dangerous to overstate than understate. And second, I'm sure I've often understated too. Sometimes what I have to say seems to me to be trite – the retelling of old proverbs. Where that's so, I'm more reassured than embarrassed, because there's comfort in company. Sometimes what I say sounds outlandish. Then I worry for a moment that I've taken a wrong turn, but when I look again at absolutely any human (or tree, or mouse, or wave) I feel that even my most baroque statements fall far short of the intoxicating polyphony of the real cosmos.

The book's chapters are arranged by theme, but every theme found its way into every chapter. Religion spills into biology, biology into death, death into perspective, and so on. That might look untidy, but it's unavoidable, and the very untidiness tends to support (doesn't it?) the notion that everything, everyone and everywhere is a mesh of edges; that edges are the stuff from which the cosmos is made. When we're thinking as proper humans should, we think in metaphor, which means that when we think about anything, we should always be thinking about (potentially) everything. Everything illuminates everything else.

The book is in four parts. The first is a chronological journey, starting with the time before there was time, and moving from the evolution of species to the evolution and decay of cities and states. The second part looks at how we spend our time in this edgy cosmos: we worship, we paint pictures, we write poems, we eat, we starve, we travel, we are altruistic and brutal. We are dissatisfied with our day-to-day modes of consciousness and try to transcend them. Thinking of ourselves as edge-people in an edgy place might shed some light on what we do, why we do it, and perhaps even what we should do. The third part is more philosophical – though you'll find plenty of real journeys there too. It asks *why* we see what we have seen at the wild frontiers visited in Parts 1 and 2. It's about the general characteristics of edginess. The fourth part is about the fictions and strategies of the centre.

Of course I hope you read the book, but if anyone puts it away and goes off to *be* edgy instead of reading yet more about edginess, I'll be happier.

This is a book about actual and metaphorical edges, and about the significance of metaphor and the importance of the actual. The real interconnection of everything means that it would be wrong to distinguish neatly between the metaphorical and the actual, and

I don't. There is, I think, no difference in substance between the edge of a continent and the edge of an idea, or between the edge of life and the edge of an artistic movement, or between the centre of London and the centre of a particular mode of consciousness. I unashamedly conflate centres of political power, centres of all kinds of orthodoxy, centres of cities and citadels of the Self. This conflation, I imagine, will be the main bone of contention. Many readers will start off thinking that the book is a compendium of category errors; that it is a schoolboy error to deal with the edge of a cliff in the same paragraph and with the same cognitive tools as the edge of a state of mind. Many readers will no doubt still think that at the end. But bear with me for a bit.

This is a book against thought, generalization and theory, and a celebration of embodiment and specificity. Proper thought, I'm sure, is what you do with the whole of your body, not just the thin neuronal rind covering the brain. Yet this is a *book*, and therefore full of thoughts rather than living. It would be far better to take soup to a homeless shelter or watch foxes with your kids than to read this (or any) book. It is also a short book, and so hooching with generalization. It is also, I'm troubled to see, an attack on theories of everything that itself turns out to be a theory of just about everything.

It is a celebration of individuation (because the more individuals there are, the more edges there are) and an assault on uniformity. It is an assertion that individuation can only be expressed properly in relationship – where the edges of one individual meet the edges of another.

It is a series of random but hopeful shots in a war against polarities, and in the service of holism. It is an expression of distrust in top-down pronouncements, and therefore an appeal for the courage to live with uncertainty. It contends that we get the best views from the edges, and so there have been many uncomfortable hours when I've thought that this is really an epistemological tract.

I'm always applauding (good) edges, even when I'm talking about transcending them, crossing them or moving on to the next one. I write a lot about the joy and the fundamental nature of move-ment, and the imperative of restlessness, but my reasons for doing so are not – or not only – Bruce Chatwin's.[1] They spring from the nature of the cosmos itself (and there's a grandiose statement if ever there was one), which, following Bergson and McGilchrist and many others, I take to be *process*: an unrolling, an unfolding; a con-stant revelation of new edges. The universe, I've already suggested, is woven from edges: it is a whole composed of abutting individuals. We can feel this texture under our feet as we walk. It is pleasurable. It is also vital for movement, for we could get no traction on a com-pletely smooth universe.

Movement – so essential to thriving – can't be measured in miles. In fact there's often an inverse relationship between the number of physical miles travelled and the amount of true travelling: the amount of real movement. Someone who stays rooted in one place and knows it intimately – seeing the turning of the seasons, the jour-neys of the snails, the passage of the birds, the springing of the leaves – has a relationship with movement and process unknown to the Frequent Traveller in the airport lounge, or even to the mission-focused hiker, striding from point to point, staring at his GPS rather than at the dew on the cobwebs. The main danger with doing all your travelling metaphorically – as we'll see when we discuss the idea of pilgrimage – is that metaphorical travel isn't embodied travel, and since we have bodies, unless our bodies go with us on our trips, the trips won't do the work they might do and need to do. But the village naturalist – who's never taken a flight, never passed her driving test and hates to be anywhere she can't hear her church bells marking the hour – is sure to take her body with her when she goes to see what her beloved beetles are doing. She's a proper traveller.

The book is not (or not consistently) an argument – though when it is, I hope it is consistent. I am tired of arguing, and argument as we usually know and use it is a device of the centrists.

Sometimes you might think I'm contending for anarchy. I'm not. Read on, if you can, and you'll find that I have some tweedy and mitred allies, and many who know who they are because they know where to put their feet in an atavistic dance. A world without Chartres is dangerous and desolate.

I espouse literal eccentricity. And literal eccentricity joyously seeds all sorts of other eccentricities, many of which are benign, and which in any event are a lot more interesting than what goes on in the centre. Which biography would you rather read: *A Life on the Edge*, or *A Life at the Centre*? More revealingly, perhaps: which one do you think would be the biography of the kinder person?

More than anything else, this book is a self-help manual for those with ontological vertigo: the sickness we feel when we look over an edge into an abyss. And that, I take it, is everyone. Or everyone thoughtful.

We tend to think of vertigo as horrible. It can be. But think of a fairground. There, people (usually children, significantly) pay good money for vertigo. They go up and down on big wheels and Waltzers, and plunge into the dark on the ghost train. They look down at the fast-approaching ground, their stomachs in their throats, and scream with fear that is also pleasurable excitement. There is comfort in the fact that they all scream together.

I would like to convince myself and others that we can live with *that* kind of ontological vertigo.

The book has a lot of me in it. I'm sorry for that, but I can't help it, because the edge on which I stand is the only place I can be, and the edge I am is the only thing I can be. And though there is plenty of hubris here, at least I don't have the hubris of pretended objectivity.

The whole is a story of surpassing strangeness: the strangeness of the everyday; an ancient and explosive story; grave and puckish; farting at our presumptions and finally judging them with terrible implacability.

If the thesis doesn't convince, be gracious, and think of the book as an extended thought experiment. How would the world look if we abandoned the illusion and conviction of stasis? If what we perceived ourselves to be was pushing always into unknown territory?

If what we had been told by the centre about what is important and true turned out to be wrong? If individuals weren't more themselves the more they defined themselves in terms of their relationships? And then perhaps wonder if that's all that different from the way things really are?

Perhaps ask too how things would be if we shed our centres, and organized ourselves and our societies like slime moulds or mycelial networks?

If all else fails, and the book doesn't even work as a thought experiment, perhaps see it as a series of dots on a canvas. Depending on where you stand, and what the light's like, and how much you screw up your eyes, you might see some forms emerge. Or (and this is my last go at a pre-emptive apologia), perhaps it's a meditation, for meditation is all about *attending*, and that's always worthwhile.

There is quite a lot in the book about death. I didn't plan it that way. But as I wrote, it seemed impossible to ignore one of the great edges – an edge that conditions our view of all other edges. To think about your own death isn't morbid. Think of it as simple ecology – as all about recycling. Think of writing that's about death as travel writing.

I would have very much preferred to write a book that simply examined and applauded edge-people, edge-places and edge-notions. I am not by nature a controversialist. I am the least political person I know. I have a visceral distaste for political discourse. Quite literally. It makes me feel ill. But this is often a political book, because politics is what happens whenever edges meet (which is always), and what happens badly whenever the centre tries to control the edges (which is always). There is, for instance, a psychopathic war on the non-human world – a war I've documented from other angles in much of my previous writing. I've tried to frame my criticisms as corollaries of my celebration of the edges. But even when being implicitly rebuked, the centre elbowed its way back towards the centre of the argument. It would far rather be criticized than ignored. And since it was there, blustering and pontificating, it had to be answered squarely.

I hope I'm not bitter or violent. But as I lived with the edges, and learned to love them all the more, I found that I could not merely praise and honour them without naming and denouncing their enemy – the centre. And so I find I have written a book that is as much about the badness of the centre as the goodness of the edges. I'm sorry for that. I'm sorry if my words are rancorous – not least because rancour is at odds with that most distinctive of edge-characteristics: unconditional kindness and generosity towards strangers, whoever they are. And not least, too, because all the centre *people* caught up in my rancour are, precisely because they are people, edge-creatures too, poised on the cusp of the same darkness, uncertainty and despair as I am: fellows in the vertigo that is a sign of being alive.

PART 1

Edging Through Time

1

Beginnings

ABOVE A BLACK ICELANDIC beach there is a dark house. It looks out over nothing – if the sea is nothing. But most of all it looks up. It feels tilted towards the sky. The sea snarls at the door, which is white and rough with salt.

I used to go there in the winter. Only in the winter. I do not know how a house like this could survive in the summer. The summer is not its time. I imagine that the house would wither in the sun.

In the daytime, such as it was, I threw stones into the sea, watched for the sail-fins of hunting orcas, read the Sagas at the kitchen table, and helped to feed the sheep. In the night, which was most of the time, they fed me on puffins, razorbills, sheep offal and shark. Then I'd put on my down jacket, grab a blanket and go outside for the real business.

The farmer had dumped an old sofa on the clifftop, and there I'd sit for as long as I could bear the cold – or as long (which was generally shorter) as my nerve held – looking at the sky. I had none of the comfort of a medieval stargazer who thought he was in an ornately decorated cathedral. I was looking out into an expanding universe. Most of what I could see was hurtling away from me: hurtling into nothing, into no place and no time – for neither space nor time had been created where it was going. Earth was getting lonelier and lonelier; all the time further away from most of what there was. Most of the light that reached me was old, dusty light: light born long before I was, which would continue travelling long after I had stopped being. I was a Copernican animal, quivering in the cold at the far end of all the worlds.

No one can take this sort of thing for long. I don't know why I did it at all, let alone why I came back year after year for more. It always ended the same way. I'd go back into the house, chastened and subdued, to watch game shows on TV with the farmer and his wife, and hope they'd reach for the bottle of brennevin.

As we watched and drank, everything rolled on outwards. I was on a part of the everything, riding the uncoiling wave, and so were you, and so we are.

Creation exploded outwards. It's not right to say creation started at a centre, for space and time were both forged in the explosion. Even if it did start from a centre, everything came to be by way of edge-formation. Creation billowed out. It unrolled; it unfolded. If there was a centre, it legislated by way of devolution to the edges.

Once time existed, things could happen in time, and the way everything happens in time is by process (not – absolutely *not* – by *progress*). Process happens at the edges, whether they are the actual edges of an expanding universe, the edges of genetic orthodoxy, the edges of an idea, or the edges of life itself.

In the medieval conception of the cosmos, beyond Saturn is the *stellatum*, the abode of the stars, whose far edge marks the limit of what we can see. Beyond that is the *Primum Mobile* – the 'first movable'.

Beyond this, at least according to Aristotle, there is no place, void or time. The *Primum Mobile*, thought by the medieval West to be powered by God (in Aristotle, by its love *of* God), rotates, and its motion powers in turn the motion of everything beneath it. In this scheme, therefore, the ultimate edge – as far away from the perceived centre of the cosmos as can be – is the engine driving earthly action.[3]

Edges were there at the beginning. First there was nothing; then there was something. There was an edge between something and nothing. The things proliferated, and so did the edges. The things differentiated, so spawning new generations of edges of different kinds. Or, if you prefer, there was chaos, and out of it came order: there were boundaries between things.

God, in the Judaeo-Christian and many other traditions, used edges as the governing motif in creation. Light and dark were separated, creating day and night. The waters were separated from the sky, and the sea from the land. In Jewish tradition, God embedded in everyday life reminders of the importance of edges, for garments are to have long, unmissable fringes: Tell the Israelites, God told Moses, that '[t]hroughout the generations to come you are to make tassels on the corners of your garments . . .'[4] The custom is preserved in the *tzitzit* – the tassels worn by Orthodox Jews.

It's probably a good idea to maintain the difference between the night and the day, for if you don't, moths get disoriented, bats starve, crops go unpollinated, corals die, clownfish eggs never hatch, baby turtles flounder onto the promenade rather than into the sea, humans get fat, depressed and irritable, and hormone-related cancers skyrocket.[5] And it's probably a good idea to maintain the difference between the sea and the dry land, because humans find it easier to live on the land.

When we first became *us*, in the Upper Palaeolithic, we were on the edge of a wholly new way of being, clutching our radical new alchemical wisdom and our dramatic new cognitive tools. Now everything could be anything else!

*

Very early on a grim, drizzling November morning, somewhere near a service station on the M5, my mother – I expect rather quietly, not wanting to make a fuss – pushed me out of her. It was my first journey, and perhaps my most consequential – though I don't know yet about my death.

I travelled over the edge of my mother's insides, through no-man's land, still umbilically attached to her, though by the end of the journey I had started to breathe air. When the cord was cut, I was catapulted over the border into the true outside. This trip was the start of me as a discrete entity. Or discrete in a few limited ways. I – or something biologically continuous with me – had existed for the previous nine months, but my biological boundaries were blurred. In some ways I had been one with my mother. Now, though – and whatever your convictions about the ensoulment of the foetus – I was, though I couldn't crawl or walk, up and running in the human race; running immediately towards death. I was unarguably an object of ethical and metaphysical interest. If you believe the Judaeo-Christian tradition, and if I wasn't made in God's image in utero, I was suddenly made in His image as a result of that journey through the cervix and along the birth canal. If I wasn't a human before, I was a human from the moment of that first gasp of foul but medicated Cheshire air. By suckling, urinating, loving, fighting, and eventually writing sentences, I was being sacramental. I couldn't help it. No one can.

We're beings in process – but, if we're wise, we repudiate progress – a clunking, mechanical word, and the favourite of those who would unmake us and put spreadsheets in our place. We are defined by our journeying – by journeys which start in and trek through liminal places. Half of me struggled up a fallopian tube to meet the other half which was coming down it, and the adventure began. That journey down the birth canal near the M5 was of a piece with my beginnings.

It's a good idea to be what you are – a traveller – rather than trying (as most of us moderns do) to be the sedentary creature you're not. Not only will trying to be sedentary kill you, it'll fill you with well-deserved neuroses while you're being swept along against your will. Swim with the stream of process and your coronary arteries and your psyche will thank you.

My family weren't at the centre of anything. My forbears came from Sicily and Ireland and Somerset and Lancashire and Wales, and from the old certainties of language, penury and community. They eked out a living doing what they could: a bit of cleaning here, a bit of soldiering and sailoring there; selling insurance door to door; working in paper mills and waiting to join the ranks of the respectable by dying properly and being buried in the same municipal cemetery as the boss.

My mother and father left their own backgrounds behind and set up together on the edge of everything other than basic decency. They were on the edge of ruin, the edge of acceptance and, most importantly for me, the edge of Sheffield.

Our suburban road ran uphill from our house to the wilderness. There, by the bus stop, the city twinkling on one side and the moors black on the other, the streetlights stopped and the howling and the killing began. I slept always with the window open, because I wanted the wild in my bedroom. It always came. There was I, tucked up, my mother playing Scott Joplin on the piano downstairs and my dad grouting the kitchen tiles, and I was breathing fox-breath and the last gasps of Bronze Age sacrifices. There was nothing more normal in the world.

I was a passionate naturalist, and every day my dad and the local kids brought me dead things to stuff and pickle and pull apart. These creatures had obviously gone over a pretty significant kind of edge.

I wasn't much good at schoolwork, but I was contrary, and interested in the inadequacies of the explanations we were given – interested, that is, in what happened at the edge of orthodoxies. I must have been hell to teach.

As a kid, it was always the edges for me. I didn't even like the middle of the day much. There were far more animals to see in the very early mornings and in the evenings – at the seams of day and night. Everyone said the day was my time, and the night was not, but I preferred the night. The day and the night were different: I liked to watch them meet. Edges gave more. You got interesting effects when things seeped into one another – from sunsets to new ideas. You got exciting sounds when things crashed into one another – when edges came together. And crashes changed things fast.

My allegiance was to the edges. So it was for a while.

For all small children of any age, everything is new, and so every moment and every place are edges of experience, peered over with dizzying excitement. Their only tribe is the tribe of everything and everyone.

There are some edges children don't see, though: they are blind to race, class and badges of status. Not until the snake offers them the fruit of the tree of tribal knowledge are their eyes opened to such things. Until then, just about everything is one, and everything is an edge. It's not a contradiction; it's one of the most basic facts.

Since every component of everything was a new vantage point, the universe was vast and pulsating with possibility and the inevitability of discovery. Everything was new when I saw it from an edge, and I didn't see anything from anywhere other than an edge. The newness itself had edges – *was* edges. What an astonishing place this is, says the edge-child about the edge-world, and ecstasy shudders through him.

Then, suddenly or gradually, edges begin to mean something very, *very* different. They begin to be important because they define a stockade; mark out a boundary; say to you that *you're inside.* Or *outside.*

We struggled to get outside our mother at the moment of our birth. We revel in our outside-ness in the early years of childhood, and then, weirdly, we struggle to get inside somewhere again, and to stay there. We're no longer content to belong to the whole cosmos; we need to belong to a tiny subset of it. We build walls around ourselves, perhaps invite a few others in, and call it security or sociation or even, if we're pompous, a society or a nation. We build gated communities, or wheedle our way into existing ones, and they become our metropolises: our centres. We are fundamentally edge-people, happy and productive only when we're on the edge, and yet we turn the edge – the glittering source of perspective and innovation and iconoclasm – into a constraining wall which *prevents* us from having perspective, or being innovative or iconoclastic.

It's a piece of diabolical inverse alchemy. Edges were everything before this Fall. Now, darkly reconfigured, they confine, restrict and define us. They tell us that we behind the wall are the centre – and ultimately that *I*, at the centre of the centre, am what's really important.

Tribes march into a child's life. Gangs at school, which teach the lessons later to be applied in the boardroom. Teams. The cohort with *those* trainers or *that* phone. The cool and the uncool. Groups whose only purpose and joy is the exclusion of non-members.

I have been totally successful in only one thing in life: resisting coolness. It's never been able to touch me. I have been relentlessly uncool. I've never had or wanted the trainers, the phone or the music of the moment. I've always been shambling, shabby and out of touch. But this, of course, is tribe-membership of a sort too. Hear the boast? Hear my contempt? My sanctimonious belief that there are things more important than phones, and my pride in my membership of the tribe with its sights set higher?

It was about the age of ten, I think, that I began to recognize and rejoice in my membership of the It's-Cool-To-Be-Uncool club. That's when I began to define myself by reference to what I was not. And when I did, the edge of me and the edges of others loomed so gigantic in my vision that I began to lose sight of the other, subtler edges which until then had been my whole experience and delight.

I remember the first hissings of the tribal snake as it slithered into my garden. They were, oddly, religious hissings. Oddly, because we weren't at all a religious family. We went to the crib service at the local Anglican church at Christmas, and that was about it. But there were some around us, I noticed, who didn't go to that service.

'Why not?' I asked.

'They're Catholics.'

'What does that mean?'

(Deep intake of breath.)

'Well . . . they worship God differently. They're still good people.' (Note that I hadn't suggested that they might not be good. *Hissssssssss*, went that snake.)

'What are *we* if we're not like them?'

'Well, we're Protestants.'

'What does that mean?'

'Good question.' (As if that answered it.)

'Well, what does it mean?'

'Just that we're not quite like them. They like candles more than us.'

'I like candles a lot. I want more candles.'

'Good. We'll get you more candles. Your fish fingers are getting cold.'

I looked it up. I was like that. Protestants, apparently, were called *Protest*ants because they protested against the way Catholics did things. I thought it was strange to give yourself a name that says that what you really are, deep down, is *not* someone else. I had a friend, Chris, who lived up our road (even nearer the wilderness than us). Wouldn't it be odd if someone asked my name, and I said, 'It is *I'm-not-Chris*'? Yet that, I discovered, was what Protestants did. I soon discovered that, odd though it might be, it was the general way of things.

The Israelites sliced off their foreskins because the Philistines didn't slice off theirs, didn't eat pigs because the Philistines did, and, right from the start of Israelite ethnogenesis in the little highland villages of Canaan, had plain pots and an egalitarian social structure because the Egyptians were ostentatious and hierarchical.[6] The

Romans defined and buttressed their identity by contrasting themselves with the effete, devious Greeks and the hairy, bloodthirsty, child-sacrificing Carthaginians.[7] 'Why do you vote GOP?' might well elicit the answer: 'Because I'm not one of those God-hating, gunbanning, foetus-killing, freedom-loathing Democrats. No, sireee, I am not.' The question 'Why do you vote Democrat?' will be met with some iteration of: 'Because, unlike *them*, I've evolved since the Iron Age.'

This malignant *othering* isn't something we do in our capacity as individuals. Individuals are kind. Othering demands surrender of our individual selves and the substitution of a corporate identity. It starts when we trade in our soul for the soul of a state, a religion, an idea or a corporation. Thucydides says it all. Why does Homer have no barbarians? he asks. He supplies the answer himself: *Because the Greeks were not yet a people.*[8]

Here, then, is a crucial distinction: between edges in the childhood sense and edges in the post-childhood, tribal sense.

Childhood-type edges are (literally) wonderful. They are of the essence of humans and of human thriving and of decent epistemology. Edges are necessary for differentiation, and good edges create the variety which, in all decent societies, is a cause of pride and joy. Post-childhood, tribal edges, on the other hand, are self-constructed or adopted walls designed to preserve a centre – whether that centre is a city, a nation, an idea or the Self itself. They are very dangerous indeed. They are the roots of denigration, self-satisfaction, toadyism, ostracism, racism and chauvinism of various kinds, epistemic blindness, greasy poles, long hours at the office in an attempt to reach the 'real centre', strategic losses to the boss at golf, and war. They make us define ourselves in terms of what we are not, which ends with us not being anything positive at all. And these edges are always far more boring than the childhood ones.

In this book I examine and honour childhood-type edges. I can't ignore completely the adult edges that delineate and protect centres, but they have had much more attention than the childhood-type edges, and I do not focus on them here. We all know what they look like. They are made of barbed wire and snubs and sneers and lots of paperwork and invitations that never arrive, and little huddles in the corner of the room, and are patrolled by sour-faced functionaries who think Faust got a good bargain.

From now on, when I talk about edges, I'll be referring, unless I say otherwise, to the childhood edges.

Before going any further I must, in these febrile times, spell out what I'm saying and not saying about diversity and nationalism.

Are things joyously one? Yes. They are also joyously many.

So, about human diversity: it is not liberal or intelligent to pretend we're all the same. It is both liberal and intelligent to assert that we are all equal in value and significance; all infinitely precious and dignified.

And about nationalism: are you a good patriot? Go for an open-eyed walk through the country you call yours. If you've done it properly, on your return you will acknowledge that there is nothing essential, nothing valuable about 'your' country that can be captured in a flag. To say otherwise is to insult 'your' country grievously. You will also see that the country is not 'yours' in the terms of your previous fulminations, and is all the more yours (in the sense worth having) for not being so.

No doubt flags have their uses. I leave it to more worldly and pragmatic commentators to spell out what they are, but I suppose they are markers of a social contract, the terms and tone of which will vary with history and temperament and certain decisions about what the limitations on freedom should be. But nobody should live or die for a social contract. People should die only for the things they live for, or which are necessary for living: family, for instance, or food or freedom. I doubt anyone really dies for their *country*. For their friends in the regiment – for an Agincourt band of brothers – yes. For honour, fear or the idea of home, yes. But for a country, no. Flags are flapping pieces of cloth. Every national parliament has one on top, but the cloth doesn't mark a place where real living happens. Parliaments are mortuaries. Proper patriots resent flags, realizing that they can't begin to describe anything about a truly beloved country; realizing and resenting the sacrilegious misrepresentation that they do.

The exhilarated child in the Sheffield garden had, for a while, the fresh, receptive, untrammelled psyche of the Upper Palaeolithic. It did thrilling things for him. So it did for us all in the bright morning of our species. Everything was one, yes: in one sense we acknowledged no distinction between us and the rest of the cosmos. We were embedded in the natural world, not striding colonially through it as we later did. But in another crucial (and complementary – and ultimately identical) sense, we saw, felt, tasted, smelt and wondered at the dazzling variegation. We looked over each of the infinite edges of things and gloried in each and all and shouted 'Yes!' to life.

We didn't have proper centres then. Instead we wandered and wondered after the caribou herds and the seasonal waves of berries. Each footfall and each glance was an edge to look or step over in the succeeding moment. We looked out of ourselves and walked away from ourselves. It's the way things are meant to be.

How do you know if you're living right – living edgily? You'll know if you're ecstatic.

Ec-stasy: standing outside yourself. If you're living all the time on the edges of things, and on the edge of yourself, it's not surprising that ec-stasy is as much a part of your diet as blackberries in the autumn. Ecstasy should be the human condition. Living at the centre of yourself isn't fun and isn't good for you. That's not a vacuous piety; it's well documented. If you want to increase your life expectancy and be happier, do things for others: pour tea in a care home; hold the hands of the dying.[9] Centres are deadly, and the ultimately deadly centre is yourself.

Being yourself properly is terribly hard. It involves being everything, and knowing that everything is shared by everyone. Here is Thomas Traherne on how to be yourself, and so how to be ec-static:

You never enjoy the world aright, till the Sea itself floweth in your veins, till you are clothed with the heavens, and crowned with the stars: and perceive yourself to be the sole heir of the whole world, and more than so, because men are in it who are every one sole heirs as well as you.[10]

There's a vast literature on the evolution of the notion of the individual,[11] and a tendency to suggest that the individual as we know it is a modern creature, born, maybe, at the Renaissance (perhaps with Pico della Mirandola as its midwife), and certainly doing much of its growing during the Reformation and its aftermath.

I can't do justice to these ideas here. I merely observe that their offence is not mainly that they patronize ancient and non-Western peoples, but that they imply that a diminished *communitarian* sense of the Self (which certainly did occur during and after later modernity) should produce a more satisfying sense of one's individuality. Surely exactly the opposite is true. Those who define themselves in terms of the nexus of relationships in which they exist are more, not less, themselves. Trawling my own experience of humans, I can't think of a single counter-example. If the growth of individualism is an incremental hardening of our individual carapaces, I for one regret it. Individuation should maximize the potential for relationship, not reduce it. What's happened since late modernity is simply that we've got lonelier, and there's less to comfort us in our isolation.

We will see in Chapter 4 that centres allure, and that this allurement can kill, as flames consume moths. But edges allure too – in good and bad ways. For every good way there is a bad way. That's how the world seems programmed. The cosmos is not dualistic in the sense that the good balances the bad – the good always wins in the end – but in the sense that everything has a flip side (though the sides are not of equal value, significance or beauty).

The edges of experience attract adventurers, poets and mystics. They also attract perverts, addicts and pornographers. The edges of the land attract painters and meditators, and the desperate fling themselves from Beachy Head. The edges of a field sometimes hum with the life and variety denied by the monocultural field itself, but may also bristle with used syringes.

Whom do we admire? The woman ahead of her time; the artist who goes beyond the conventions; the altruist who goes beyond the call of duty. Whom do we despise? The man who runs ahead of his ability; the sensualist who flouts rightful taboos; the selfish bastard who tramples others regardless of censure. We applaud the jazz pianist for improvising around a Bach fugue; we frown on the jazz trumpeter whose take on the same fugue is incoherent and narcissistic noise. We approve of great wealth when it is won and used well, but not of the same wealth won by exploitation and used for self-aggrandisement. We are moved by the dignity with which some die, and appalled when someone dies screaming and cursing.

For good and for bad, we pay special attention to edges. We can't help it: that's how we're made. More work goes into a herbaceous border than a lawn. Even if we're not Orthodox Jews, we pay particular attention to the fringes of our garments (think of all that lace). We rush out into the cold to look at a sunset. My camera even has a special sunset mode. And as we have seen, and will see further, God too is very concerned with edges.

I recently returned from Normandy. I went to see the Bayeux Tapestry, but barely noticed the epic story told by the bulk of the tapestry because the borders were so fascinating – prowled by fanciful beasts. At the moment, much of my time is spent with medieval illuminated manuscripts. The body of the manuscript, which the monks are supposed to have thought was the real point of their work, is generally superb, but the real interest (one suspects for the monks as well as for me) is in the margins. There, on the edges, is the real and imaginary life of the Middle Ages: wooing, wood-chopping, belching. weeping, hunting, pig-killing, disembowelling, praying and dragon-killing.

We mark our edges. Boundary disputes are the most bitterly

contested type of litigation. Litigants beggar themselves to establish the right to a six-inch concrete strip next to the septic tank. Nations blithely send their young to die for the equivalent of that strip. We all wear uniforms – some on our bodies, some on our faces and in our histories – declaring our entitlement to something and, more gratifyingly, another's non-entitlement. Monks mark the boundaries between the times of day by ringing bells and reciting the daily offices. The muezzin in the minaret does so for everyone in the district. Factory workers clock in and out, and their productive periods may be despotically distinguished from their less productive by an all-seeing program.

Most of the old divisions of the year (the solstices, the quarter days, the equinoxes, the great religious festivals) have been forgotten in the West, but we still buy chocolate eggs at Easter, eat too much at Christmas, and count down the seconds to the New Year. We still recognize the significance of the moment when a child is expelled from its mother, and dutifully turn up to bury or burn the friends and family who have travelled over the far edge. In our gloomier and more honest moments, we realize that we're always swaying drunkenly on the edge of eternity.

We're edge-people in an edge-world. Good living starts with knowing this, and is about partying on the lip of the precipice – or at least not being too terrified by the view – and about seeking out good rather than bad edges.

Guillemots and razorbills nest on huge, exposed cliffs. Space is always at a premium, and success in life depends on acquiring and maintaining a good position on a ledge. One inelegant shuffle and your egg – your posterity – will fall hundreds of feet into the hungry sea. That's us too. Except that we're on the ledge all year, every year, for however long we live, and the best way of occupying the ledge isn't to peck your neighbour, but feed her. There's another lesson here. It's not necessarily the ledge – or edge – that is good or bad, but our behaviour when we're on it.

2

Everything Evolves

Humans have evolved to their relatively high state by retaining the immature characteristics of their ancestors. Humans are the most advanced of mammals – although a case could be made for the dolphins – because they seldom grow up. Behavioral traits such as curiosity about the world, flexibility of response, and playfulness are common to practically all young mammals but are usually rapidly lost with the onset of maturity in all but humans. Humanity has advanced, when it has advanced, not because it has been sober, responsible and cautious, but because it has been playful, rebellious, and immature.

TOM ROBBINS,
Still Life with Woodpecker[1]

WE'RE NOT ONLY DESIGNED *for* edges; we're designed *by* them.
A hundred miles from the Scottish mainland, slabs of green water smash into the most important seabird cliffs in the world. The sea is notoriously dangerous here. It sucks up boats and boulders like spaghetti. The waves can be the size of houses. It is often impossible to land, even at the modern jetty in a boat with all the technological bells and whistles. These islands – the St Kilda archipelago – were occupied by humans from at least the Neolithic until they were evacuated in 1930, and during all that time humans barely fished. It was just too risky. They ate seabirds – mostly fulmars and gannets – and their eggs instead.

St Kilda is a masterclass in vulnerability, and hence of evolutionary innovation and (since Darwin's rule is blind and impartial) evolutionary failure.

There are strange buildings here – stone rings nearly 4,000 years old, and stone structures with curved forecourts like the horns of Highland cattle. Nobody knows what they were for. There is nothing like them anywhere else. So far as we know, no influences passed to or from them. They were a uniquely local solution to a problem that was doubtless not unique to St Kilda. There is now no way of telling whether their innovation was good or bad.

Nothing in biology is good or bad, except the context makes it so. We can argue about whether the same is true for human culture. The St Kilda architects might, for good or bad, have revolutionized the Neolithic and changed the course of our history, had those slabs of water not denied them the chance.

The same is true for the now-extinct St Kilda house mouse. Before 1930, by the light of candles made from the burning bodies of Leach's Petrels, the islanders watched these mice gnawing the puffin bones left from the evening meal. There are now none to watch. The only St Kilda house mice nestle amongst the mothballs in museum drawers, or swim in bottles of formalin. They were part of an ecosystem of which humans were an essential part, and were extinct within two years of the evacuation. What might they have been had they stowed away in one of the boats that laboured biliously to the mainland? They might have outgunned all the house mice in the Kyle of Lochalsh, then in Mallaig, then in Edinburgh, then in London, and then in the world.

St Kilda *field* mice, though, survived the departure of humans, and ballooned. They now weigh twice as much as mainland field mice. Their diet has changed too. They used to be mainly vegetarians, but now dead seabirds are important. Natural selection, very fast, has fashioned giant, carnivorous mice.

It would not have done so – or at least not nearly as fast – had St Kilda not been an island archipelago. The field mice illustrate the 'island syndrome' – the principle that, on islands, big things tend to get smaller, and small things tend to get bigger. St Kilda hatched mammoth mice; Cyprus, tiny elephants and hippos. Put a pigeon on a relatively predator-free island for long enough and it will become a dodo.

The reasons are much debated and do not concern us here.[2] The point is that edges are creative, and often vigorously so. On the Wolf and Darwin islands (and only there) in the Galápagos archipelago, vampire ground finches, particularly in hard times, tap and drink the blood of blue-footed and Nazca boobies. Hawaii has a population of *Ariamnes* spiders which, over the last 1.7 million years, have become night-hunters of other spiders.[3]

To watch evolution at work, we should go to the edges: of a petri dish, where bacteria meet the big wide world; of landmasses, where the challenges are different from those far inland; and of genetic orthodoxy, where natural selection has something new to work with.

Central populations are stable. There, often for millions of years, the détente between organisms and their places has been fairly fixed. The conversation between genes and environment has become polite, predictable, and often inaudible. If there are no new threats or opportunities, there is nothing to discuss.

Evolutionary discussions are conducted in two languages. There is the conventional language of genetic mutation, in which natural selection assesses whether or not a mutation confers a selective advantage. And there is epigenetic language, in which the conversation between genes and environment may itself affect the switching on and off of genes in a heritable way. Epigenetics has brought the French biologist Jean-Baptiste Lamarck (1744–1829), long reviled for his belief in the inheritance of acquired characteristics, back into the biological fold. He is not just tolerated; he is in the vanguard. He has tenure and lots of grant money.

Whichever language is used, it is one of the edge-dialects.

In the conventional analysis, a mutation is, by definition, something out of the usual order of things – far from the middle of any bell curve invoked to describe that gene's behaviour. Epigenetics, too, selects for an innovative mode of gene behaviour at or over the edge of the environmental conditions that characterized the previous status quo. Evolutionary change is about the endorsement by the environment of edge-organisms and edge-behaviour. It has nothing at first to do with the centre – though if an innovation pioneered on the edge shows its mettle, it might rearrange the centre and become (until circumstances change) the new orthodoxy.

Relationality is at the core of biology. This should be obvious. Sexually reproducing organisms reproduce by establishing a relationship in which, by the fusion of gametes to form a zygote, the individual edges of each party are transcended to create a genetically novel organism. One set of edges has to come into contact with another set. One edgy boy has to meet an edgy girl. Even in asexually reproducing organisms there is an obvious relationship with the environment, which will be watching hawk-like to see if an error in the cloning or budding process produces a novelty worth promoting.

Humans, like everything else, are evolving. Since we became anatomically modern (say around 300,000 years ago), our culture has evolved more dramatically than our anatomy. There have been bigger changes in our buildings than in our bones.

Matt Ridley's views about climate change might be unusual amongst the scientifically literate, but his views on cultural evolution are far from sacrilegious, and are increasingly being assimilated into the mainstream.[4] Biological evolution, he argues, is just one special example of a 'General Theory of Natural Selection'. Natural selection, in precisely the way it operates to produce biological change, changes other systems too, including morality, religion, culture, music, politics, and so on – often, he contends, through the medium of the meme. It is the ultimate Theory of Everything. Ideas have sex with each other to produce baby ideas, and we choose which babies survive. It is just a grand way of talking about the process of trial and error.[5]

There is no room in this scheme for any top-down revelation. It doesn't need any deistic 'sky hooks', so memorably derided by Daniel Dennett, used for constructing buildings from the top downwards.

Ridley's loathing of religion diminishes the force of his argument, but nonetheless the gist is compelling. It is supported by the observation that in every field (for better or worse) we build on the basis of what went before.

Ridley calls a host of witnesses. Gods evolve from petulant, impulsive tyrants to benevolent, bodiless spirits. Governments start as protection rackets. Language changes, without any committee deciding how it should, until it has elaborate rules. Some of those rules are invisible: if we use a word a lot, for instance, it becomes shorter, and long words change their meaning much more readily than short ones.

Ridley overstates, and some of his assertions have a dated, Dawkinsian feel about them. For him, for instance, altruism and cooperation are really just disguised selfishness. Competition is given credit for artistic and ethical creativity it does not possess. Yet the core of his case accords with what we know both about the mechanics of biological evolution and the dialectic of cultural change.

Sometimes something new comes out of the blue – a phenomenon we will look at later, known as a 'black swan'. Evolutionary theory has for a long time recognized that change can happen in big jumps (the notion of punctuated equilibrium). The fact that we as a species are woefully maladapted to the flight into our lives of flocks of black swans simply means that natural selection is not infallible. No one in the know ever said it was.

So: centres have nothing to contribute to biological innovation. Innovation happens at the edges – of established genetic make-up and, typically, of populations, where populations meet and rise to new challenges. The same is substantially true of cultural and other phenomena too. This book contains many examples. But biology gives further emphatic confirmation of our natural status as edge-animals.

Those austere Victorian schoolmasters were right – though for the wrong reasons: it is good (up to a point) to be stressed, cold and generally wretched. Biologically blessed are you when men dunk you in icy water, flog you up and down mountains, and expose you to filth and infection. Blessed are those who shiver. Blessed are those who hunger. Woe to you when men wine you, dine you, and wrap you in sterile cotton wool. It's not surprising. We haven't evolved to slump on armchairs in centrally heated rooms, eating three large meals a day. It's unnatural. It's fatal. Comfort kills.

It is literally depressing that these observations need to be made. For modern Western ways of being contribute significantly to the incidence and severity of depression – a dangerous disease with a very high mortality rate. The risk, severity and duration of depression can be mitigated by exercise – particularly outdoor exercise in green places (which reminds us of our entanglement with the more-than-human world) – by other forms of 'good' stress, and by the relationality for which our massive brains evolved. Good stresses all demand a response from us – they move us away from the toxic places we have chosen; away from the centres in which we have pitched our camps and cherished our delusions of stability and certainty.

Exposure to cold seems to sedate the black dog of depression. It activates the sympathetic nervous system and produces a gush of endorphins and noradrenaline. Since there is a high density of cold receptors in the skin, that cold shower bombards the brain with impulses from peripheral nerve endings, which might help to jerk the mind out of its depressive rut.[6] Subjects who took cold showers for ninety days had a 29 per cent reduction in sickness-related absences from work compared to those who did not.[7]

A review of studies on hydrotherapy – which included both cold and hot water immersion – concluded:

> Based on available literature, this review suggests that hydrotherapy was widely used to improve immunity and for the management of pain, [chronic heart failure], [myocardial infarction], chronic obstructive pulmonary diseases, asthma, [Parkinson's Disease], [Ankylosing Spondylitis], [Rheumatoid Arthritis], [Osteoarthritis of the knee], [fibromyalgia

syndrome], anorectal disorders, fatigue, anxiety, obesity, hyper-cholesterolemia, hyperthermia, labor, etc.[8]

It is quite a list.

If you heat yourself up to unphysiological temperatures in a sauna, it's likely to reduce morbidity and extend life expectancy. Fasting seems to improve blood sugar control, reduce the risk of type 2 diabetes, damp down inflammation, improve blood pressure and cholesterol and triglyceride levels, reduce the risk of neurodegenerative disorders, boost cognitive performance, reduce the risk of cancers and increase the efficacy of cancer therapy. As well as making you thinner.[9]

In fact just about any kind of stress, other than the stresses associated with living comfortably, meeting the deadlines of projects you know don't ultimately matter, and being sycophantic to moronic bosses, is good for you. There's even a word for the phenomenon: hormesis.[10] Roughly: whatever doesn't kill you makes you stronger. That includes being irradiated,[11] and poisoned with a wide variety of substances.[12] No consistent capitalist should be surprised. That, after all, is the fundamental assumption of economic policy in the dog-eat-dog Westernized world, where it is assumed that, overall, competition is healthy for a society. Move out to the margins, where you can live longer, more abundantly, more bracingly and more interestingly, say Darwin, our white blood cells and our metabolic pathways.

Hormesis applies to individuals, but one of the fundamental axioms of evolutionary biology is that stresses faced by the individuals in a population can drive changes to that population. Take, for example, the bacteria causing your chest infection. Your doctor gives you seven days of antibiotics. After three days, you feel a lot better and stop taking the drugs. Many of the bacteria will already have been killed. That is why you feel better. But some, with a degree of genetic resistance to the antibiotics, will have survived. If you had taken the full week's course, even these relatively resistant bacteria would have been killed – by the antibiotics, your natural immunity or both. But because they have been allowed to survive, they will form the nucleus of a whole population of relatively resistant organisms.[13]

The general point, again, is that only on the edge of the usual set of conditions (an edge such as an unusual concentration of antibiotic) do we see significant changes in the constitution of populations. In biology, as everywhere else, necessity is the mother of invention. 'Crises give birth to moguls', said the Greek shipping magnate Aristotle Onassis, whose coffers were massively swollen when the Suez Canal was closed to shipping for six months, creating a great demand for vessels to chug all the way round the Cape of Good Hope. 'The best deals and the best sex happen outside of boundaries', he went on. That is a neat summary of evolutionary theory. 'The best deals' equate to the greatest evolutionary fitness – conventionally understood in terms of the number of viable offspring. And the 'best sex'? Well, that is more complicated.

Sex is a strange business. It is a device which is somehow, and mysteriously, smiled on by natural selection, for ensuring that your offspring are unlike you. Just why natural selection should have favoured a system that produces organisms *unlike* a successfully reproducing organism – one that has demonstrated its evolutionary fitness – is unclear.[14] Sex, anyway, is the business of creating difference; of multiplying edges. It is a special and dramatic illustration of the cosmos's general tendency to individuation. It exists to generate edges, and it happens, by definition, *at* and *over* the edge of oneself. Differentiation happens more powerfully when very different individuals mate. And that happens particularly at the edges of populations and cultures. Liminal places are good at curating meetings of genetically different organisms.

Take airports.

Pete is a Californian surfer from a secular Latino background. He went to Indonesia in search of a legendary wave. The wave never showed up, and so after a week of disappointing surfing and overpriced beach-bar cocktails, he changed his flight and started to head home.

He had a long layover and a bad hangover at Bangkok airport, and fell asleep on the floor, his face covered with a towel. He woke when the towel was pulled away. It wasn't airport security but Maria, a Filipino nurse. He sat up and asked her what was the matter. She apologized, saying that she hadn't seen his chest moving, and wondered if he was all right.

Pete was touched. He was also attracted. He thanked her and suggested a cup of coffee.

'She wasn't at all my usual kind of woman,' he explained. 'She spoke stuttering English, hated the sea, loved the Virgin Mary and was a virgin herself, came from a family of small farmers, and wanted an inner-city apartment, a small dog in a coat and a dishwasher. She'd been working at a Catholic mission for street kids.'

Maria agreed that they were completely incompatible. Pete was just the sort of man her mother and her priest had warned her about. And she couldn't stand beards. If a man had to have a beard, it shouldn't be matted with salt and sriracha.

They talked for five hours, missed their flights, and spent most of the next week in bed in a doss house over a Chinese laundry.

They've been married thirty years, live in Berlin and have a dishwasher, no dog, and four children (one an expert on tapeworms, one an acupuncturist, one a scuba-diving instructor, and one a multimillionaire running a nail-bar empire in Philadelphia). Pete is clean-shaven and a lay member of the Society of St Vincent de Paul; Maria an ultra-marathon runner, a Soto Zen teacher and the manager of a shop selling drones. They already have three grandchildren. Another's on the way.[15]

Darwin would be proud. They are *fit*. Their offspring are textbook examples of hybrid vigour, to match the vigour of the offspring of mortals and gods or sea nymphs. That's what happens when edges are crossed.

We're full of inhibitions. Many of them are thoroughly good and healthy, but they do restrict possibility. Some are neither good nor healthy, and also restrict possibility. But our inhibitions behave curiously. When we cross one boundary – taking us out of our usual neighbourhood – the journey seems to dissolve other boundaries too. It's as if edges come as one bundle: cross one, cross them all. We seem to think that the usual rules governing our lives cease to apply – even to the extent of believing that, by being on an adventure, we're magically immune to disaster and disease.

Sex, again, is a good example. Sexual disinhibition is more common on vacation than at home. One of the leading studies (of young American college students on spring break) showed that 30 per cent of the men and 31 per cent of the women had had sex with a new partner met during the holiday.[16] The authors of the study even coined a polite euphemism – 'situational disinhibition' – for the unholy trinity of casual sexual encounters, substance abuse and holiday sun. A majority of gay men on Mardi Gras vacations in New Orleans said that finding a new sexual partner was part of what made a vacation fun, and a significant minority believed that their interest in sex and their sexual activity increased on holiday. About half of international travellers have casual sexual experiences while on the road.[17]

Risky behaviour is commoner too.[18] UK sexual health clinics are busier after the holiday period.[19] Gay men on vacation on the East Coast of the US had unprotected intercourse with 11 times more non-main partners compared to at home.[20] Nearly half had sex with a partner of unknown HIV status, and a similar number did not disclose their own HIV status to all their sexual partners.[21]

The fact that sexual disinhibition occurs so consistently when we're away from home suggests that it stems from deep places within us. As for the risk-taking: well, evolutionary fortune apparently favours the bold sufficiently often for it to be worthwhile. This is particularly likely to be the case where the risky behaviour could, in some contexts, lead directly to conception. Here too, then, edge-crossing is a Darwinian strategy.

The restlessness seen in our holidaying, our dog-walking, our jogging, the strange static-cycling in our bedrooms, and our hopping between books, ideas, TV programmes, websites, partners and enthusiasms is a bequest from our hunter–gatherer past and an indication of our fundamentally hunter–gatherer present. In hunter–gatherer communities all – but particularly males (who tend to be hunters rather than gatherers, though women hunted too[22]) – have to travel long distances and be competent navigators. Good travellers are therefore good meat-providers, and accordingly more attractive to women. In modern hunter–gatherer societies the best hunters (and therefore the best travellers, with the itchiest feet) have more wives and more extramarital affairs.[23]

Risk-taking might often pay off, and be attractive in its own right. We have seen some examples of risk-taking on vacation, but there are even more blatant ones: leaping off bridges attached to rubber bands, or visiting war zones and then boasting about it in the bar.[24] We can go straight from ancestral mammoth-hunting, to modern gazelle-spearers in sub-Saharan Africa, to adventure and disaster tourism, to the clap clinics of Wolverhampton. The context may change; humans haven't changed much.

Other edge-phenomena from our hunter–gatherer past continue to condition our modern edginess. We have, it has been observed, 'deep adaptive responses to the natural environment', which attract us in particular to locations on a 'prominence overlooking a body of water or vista where prey may be easily observed.'[25] We like, that is, to have a room with a view; an edge-location. Every estate agent and hotelier knows the premium people will pay for a vista.

Living, loving, dying, evolving: in the sauna, the airport, the island, the singles club, at the end of a bungee cord and an umbilical cord. We do it all most vigorously on the edges.

Life began there. Story began there. And we began there.

3

A Human at the Edge of Europe

Frisia, through its position in the liminal zones of the two spheres [of the Viking and Frankish worlds] *becomes a depot for wealth and cultural negotiation. Nevertheless, it remains between two centres and as such does not become the central focus of either sphere with which it is connected. Liminality, not belonging to one sphere completely, but to two, sometimes shifting spheres simultaneously, creates opportunities for connectivity with other areas, resulting in cultural links, shared material culture and textual traditions. Through its connectivity with the Viking North Sea world stretching from Scandinavia to the British Isles, and with the Continent, Frisia was central because it was liminal.*

NELLEKE IJSSENNAGGER,
Central because Liminal: Frisia in a Viking Age North Sea World[1]

I AM SITTING AT THE edge of a country, at the edge of a continent. Over the sea, behind the gathering storm, is Africa, where humans first happened. Humans swept up and down and out of Africa, and as they moved, new things gestated inside and erupted out of them into the world.

The big migration into Europe happened from around 100,000 years ago. Perhaps later. But just down the road from me something very strange has been found. It is the skull of a human male. It is *Homo sapiens* all right – *us* – but with some archaic features. It is at least 210,000 years old. He was here at least 110,000 years before anyone else. He's not supposed to be here. He was in a cave by the sea.

I've taken to going there. I'll go there later today. He's become an obsession.

It's not so easy to get there. You drive twenty minutes from the house, through the olive groves, winding round the big black snakes basking on the road, along the pass between the mountains, mobbed by ravens, through the little town – slaloming, because nobody will break off their mid-road conversations just to avoid being killed – then turn steeply downhill towards the sea and on to a dirt road past a sleek hotel boasting of its 'intuitively authentic wining and dining'. You can't drive beyond that in even the most lavishly insured hire car.

Then it's an hour's walk. There's more rock than grass. Sometimes the rocks have been piled up by geology, sometimes by farmers to say that a patch of burnt hill is theirs, or to give the goats somewhere to hide from sun and storm. High up above you, away from the sea, there's a severe white ridge that looks like a fort and often has been. There are hives in a meadow, and the bees feed mainly on a type of anemone the old Greeks say presages early death. You drop down through unloved olives and little bitter oaks with acorns the size of plums. All the time there's the sea – the great fact. If it's calm there'll be a boat rocking out there, pulling in little red fish.

The rocks have been bored and carved. Every one is a skull with mis-set eyes. Everywhere in southern Greece appraises and interrogates, and here it's a really aggressive cross-examination.

Soon the track gives up, and you turn straight towards the sea, through olives bent halfway to the ground by the wind from the west. They are nearly flat, like roof tiles. The goats have pushed dark tunnels into the thorn bushes. I don't know how. If you stood on one of those thorns, it would go straight through your boot and your foot and out the other side.

Then the sea is suddenly all there is. It churns and booms and if you fell two hundred feet into the maw you'd be well chewed, and later vomited into the first European's cave for the gulls to finish.

29

The cave complex is off to the left, far down, just above the sea. It's dangerous to try to reach it from above, and I've only been there once, jumping off the cliff with my nerveless sons and letting the waves spit us inside. I can see into his cave from the goat track, though. It was much higher above the sea in his day, and the sea may have been a fair way off. The sea was lower then than it is now, for lots of its water was locked up in ice. The bay half a mile to the north, where I swim morning and night with the turtles, was a lush valley, fed by the river that runs today for about three torrential days a year down the ravine where we see eagle owls on our way to the mountain monastery. The man would have speared wild goats in the valley and brought them back on his shoulder.

No one knows if he lived here or was just picked clean here. If he didn't live right here, it wasn't far away. You wouldn't carry a human corpse far down these cliffs.

This first European was black.[2] He was a brilliant naturalist. He knew how the seasons rolled: when the birds would arrive from Africa, as he had, and when they would return as he never did; when the fish would come; what leaves would take away his pain and what root would show him his dead father. I can barely tie my own shoelaces, but he clothed and fed himself from the wild of which he and I are both a part. He must have had family, and perhaps a clan, but there weren't many humans here. All day and night he heard the sea. Whenever he turned west, he saw it. The sea was one of the many edges he had crossed, and on which he lived, and on which he built his story. An edge is a strange place to build anything, you think? There's nowhere else.

You can't see Africa from here, but you can sometimes smell it – right on the edge of your nostrils, for it is only on the edge of olfaction that you smell anything interesting, just as it is only on the edge of vision that you see anything worth seeing. Sometimes, when the oregano has withered, and the wind is from the south and the sea like syrup, there is a stewed vegetableness in the air. It comes down the Nile, spills out into the delta and streams past Crete into this bay. Sometimes Africa is sticky on the car.

I admire this man. I have no idea what he was doing here. Nor does anyone. He's the ultimate edge-person: on the edge of geography, his species, history, imagination; and the edge of everything except mainstream archaeological theory (he's well over the edge of that). If you're looking for the prototypic modern Westerner, here he is: sitting on the shore five thousand miles from base, eating goat, watching rhino, and laughing at our categories. He's the normative one. We're all edge-people, as we'll find out sooner or later, and we're a lot better off if we don't pretend otherwise.

We don't know what became of this man. We don't know if he had a partner; if some of his DNA is in our cells. Perhaps he was on his own, having arrived by some freak accident, and died thinking wistfully of Africa. But long after his death he was joined by other Africans, who probably loped in through the Levant and Anatolia, shivering in the unaccustomed cold, but wrapping themselves in animal skins, fashioning shoes to keep snow from between their toes, carrying fire wrapped in moss, and sitting for long nights staring into the flames with hot faces and cold backs. Hot face: cold back. That *is* the human condition. It was and will be so until death is abolished.

In the forest behind them shone the eyes of things that wanted to eat them. It was here, where the light met the dark and the warmth met the cold, on the verge of viability, at the edge of the ice, that we (I'll call them *we* now), brandishing our newly ignited consciousness, whiled away the night and tried to exorcize the spectres of extinction and meaninglessness by telling stories; by insisting that we *meant* something. Culture – which in its modern sense started to sprout as soon as we had modern minds – now flowered exuberantly. Notre Dame de Paris, Shakespeare and the *Mona Lisa* were conceived around the Pleistocene campfire.

Depending on how you do the calculation, between 85 and 95 per cent of our time as behaviourally modern humans (since the Upper Palaeolithic, about 45,000 years ago) has been spent as hunter–gatherers. This means not only that our childhood as a species has been spent as hunter–gatherers, but that most of our life has been spent that way. That's what we really are. Rip the suit from the New York head-hunter and you'll find an Upper Palaeolithic caribou-hunter. (It's not the focus of this book, but I suspect that many of our modern illnesses of mind, spirit, body and body politic result from locking up the caveman in us. There's lots of nonsense talked about the Palaeolithic diet, but lots of sense that could and should be talked about the Palaeolithic mental and spiritual diet.)

That's it, really. That's the history of Europe. Or at least the history of Europeans. The rest is window-dressing. Armies? Kings? Empires? Cities? Nation states? All footnotes at best. Wars are fought by hunter–gatherers behaving uncharacteristically. Kings are clan chiefs who've forgotten the rules. Empires a miscalculation of the boundaries of foraging territories. Cities, middens. Nation states, purely imaginary.

Nonetheless, some interesting and enlightening things have happened since the end of the Upper Palaeolithic. It's been a thrilling journey, as it's bound to be when edge-people are on the road. We'll stop off at some of their caravanserais: religion, poverty, death, art and politics amongst them. But again and again we'll see that wherever anything worthwhile happens, it happens where it happened at the start: at the edge. By the campfire. With the chill at the back and the slavering jaws waiting.

There's a modern campfire a few hours' drive from the first European's cave.

It's in a Greek port, next to where the boats moor. I don't want to name the port, let alone the place, for this is fragile. It is an old-fashioned café-cum-taverna. They'll give you a litre of home-made retsina in a monkey can for breakfast and sweet dark coffee, half sludge, at midnight. Whatever they give you, it comes with a small fish with big bulgy eyes, and a lump of blood sausage.

Only retired sailors and I come here. Even in the midsummer heat they wear woollen caps pulled over their ears. When, once every couple of hours, they stagger up from their table (*their* table, mind: this is territorial, and it would be dangerous to sit in the wrong place) to piss out the wine, they stagger with the kind of roll you need to stay upright when the wind from Libya is threatening to smash you into Cape Tenaro. On their wrists, lapels and round their necks are the charms of maritime saints, for even when you're ashore Poseidon can rear up to crush you. Stella Maris hangs over the door. The paint's yellow with smoke, and the walls are frilly with the fringes of the Peloponnese and the Dodecanese on maritime charts. The owner's dead parents preside from silver frames over the fridge, next to St George, who is killing a sea monster. There are full ashtrays under every no smoking sign.

I usually sit under a long-legged girl from the fifties who's showing her legs to sell ouzo; a fuzzy photo of a lamb on a spit, taken in black and white one Easter after a gallon of red wine; a salt-crusted oil painting done by a sentimental sailor of his church back home, the sea just visible; a row of brass barometers saying there's a storm heading this way from Constantinople; and clocks telling the time in Alexandria *circa* 1965.

I sit here so I can see the ships outside, and see too the painting that fills most of the wall opposite. It shows an archipelago somewhere in the eastern Aegean, with the depths around the islands rendered carefully in different shades of blue and, on another panel, a dinghy on its way to a beach.

Men (it's only ever men who come here to drink) sometimes die here, says the brisk but not unsympathetic waitress. It's not surprising. I once came in at 4 a.m. and left at 6 p.m., and three of the men were there all that time, moving only to lift their glasses to their lips, signal for more, and roll off to the urinal. They're unlikely to die anywhere else. They're not anywhere else long enough. The wheelchair ramp outside is used only to slide dead customers into the street. Everyone alive is too proud to use it.

From time to time there are other visitors: the Albanian prostitute who brings in her child first thing in the morning after a night's work, and does the rounds of the tables holding out her hand for coins but is unable to look you appealingly in the eye; the lottery-ticket seller, probably Syrian, who brought only two limbs with him on the little boat.

This place won't survive long. It tastes of too much to hold out against the tsunami of blandness whose approaching rumble drowns out even the clang of the ships outside.

Here there are many endings and perhaps, too, some beginnings – though some of the beginnings are just the stories of the worms that will eat the old sailors.

Some things, not at all mawkish, started here for me. A happy expedition to a pistachio farm. A search for the lost church of Yiannis the Theologian. A doomed chase round the Aegean in pursuit of a Dutch nymph I'd met in Sidon. A family trip, many years later, to buy black market trainers. And this book.

For so many edges came together here that even I couldn't fail to notice. These old men had spent their lives riding from one edge to another – one wave crest to the trough and back up and back down – and now they were about to cross the biggest edge of all. I marvel at them. They may seem imprisoned by memory and mortality, frozen by alcohol and arthritis, but, driven before the edges of the wind and the sea and time, they are free.

Here in the café, the sea cave, and round the Ice Age campfire is the real history of Europe. Not in the museums or the textbooks, let alone in the parliaments or the government offices. But history, of course, is written by the victors – or by the people who proclaim themselves the victors. They live in cities.

Let's have a look at them.

4

Can Anything Good Come from a City?

Over the greater part of history, the village and the countryside remained a constant reservoir of fresh life, constrained indeed by the ancestral patterns of behaviour that had helped make man human, but with a sense of both human limitations and human possibilities ... Even if whole urban populations were destroyed, more than nine tenths of the human race still remained outside the circle of destruction. Today this factor of safety has gone: the metropolitan explosion has carried both the ideological and the chemical poisons of the metropolis to every part of the earth; and the final damage may be irreversible.

LEWIS MUMFORD,
The City in History[1]

London only stimulates, it cannot sustain ...

E. M. FORSTER,
Howards End[2]

MY BALD-TYRED BUS SPLUTTERED and lurched out of Elazig, showering a date-stall with dust and maiming a dog. When I'd climbed aboard, all conversation stopped and every eye swivelled towards me. Now not even the children watched. I'd breathed nothing but cigarette smoke and the fumes of pirated perfume for nine hours.

I'd thought that Elazig was in south-eastern Anatolia. 'Nonsense,' said a very learned historian. 'It's Mesopotamia. That's important.'

The two great rivers – the Tigris and the Euphrates – arise near here. Trying to be smart, I'd said that I supposed it depended which bank of each river you were on. If you were on the eastern bank of the Tigris – the easternmost of the two – you wouldn't be *meso potamos* – between the rivers – would you? 'Nonsense again,' he'd said. 'These rivers rule the land for a good couple of hundred miles west. You can feel their pull.'

He's not given to romantic ellipse, this man, and I wondered what he meant.

The bus disgorged me by a derelict abattoir. I sat under a palm tree, smooched by cats whose fur flickered with fleas, and looked at the map. By this point, not far from the Syrian border, the Tigris – conceived by ice and storm in the Taurus mountains, gestated in Lake Hazar, arrested by dams and corrupted with heavy metals from incontinent mines – is sullen.

The river was a mile from the road, past signs to nowhere pocked with bullet holes and swinging in a hot wind thick with flies. The cats followed for a hundred yards and then gave up.

I had to give up too. This was as far along the course of the Tigris as any sensible insurer would let me go.

It was late. The last of the light was draining into the Syrian sand, but I could see that the river was beginning to be itself, not running but sidling through a plain of grit, goat dung and potsherds. I waded in up to my waist. Fish nudged me. I dried, unrolled my sleeping bag, boiled up some tea and fell asleep listening to the jackals – which have always been my friends.

The water running past my head when I dropped off was crossing into Iraq when the sun came up. In six hundred miles the river would join its sibling, the Euphrates, conceived too in the Taurus, and together they would swell the Shatt al-Arab before spilling into the Gulf in a muddy cloud visible from the moon. The siblings incestuously spawned cities.

Did I feel the tug of the rivers, as my friend said I would? Who knows what is real and what is fancy? I felt no tug, but I think I know what he meant. The Tigris was not a merry, singing river, or a solemn, mellifluous river. It was a river stately with age and influence; a disparager of less august streams. It knew what it had done to humans. It stank of old metropolitan shame; shame that rose like mist.

The river has done terrible things.

When we stopped hunting and gathering – when we stopped being ourselves – we conglomerated – notably in the cities of Mesopotamia – Eridu, Ur and Uruk, around 7500 BCE – just downstream from that camp of mine in the far east of Turkey. The reasons for the conglomeration are much discussed. The fecund mud of the two rivers is not the whole answer. Probably the cities congealed around temples designed to hold the ring in the struggle between the sea, the rivers and the sun. Though they are the ancestors of the places where most of us live today, we should not look at them sentimentally. They were tyrannous, pestilential and hellish hot. If you look at them with neoliberal eyes they were also extremely productive. Supply and demand entered into the coalition we think of today as sovereign. The coalition flogged its slaves to death. Humans lived in these cities in ways they had never lived before. Jericho – the oldest continuously inhabited settlement on the planet – was one thing. The vast Mesopotamian cities were quite another. They were on a different scale altogether, and a difference in scale became a difference in kind. Humans were rewired there even more thoroughly than AI threatens to rewire us today. There humans became commodities, to be weighed with the same measures as barley.

Language, previously used for telling creation myths, making love and giving commands about how to spear bison and find nuts, was now used to describe yields, debts and other legal obligations. Writing was probably invented in these Mesopotamian cities – not for poems but for accounting. Language had been a live, fluid medium, given to improvisation and prone to elicit immediate responses:

guffaws and tears. But when it was written down – inscribed on a clay tablet – it died. You can't riff around a tax return. It says what it says. It can't be appealed to. These dead cuneiform wedges ruled absolutely as live language never did. They represented abstractions – profits, losses, times for payment – and these abstractions were sovereign: they trumped flesh, blood and desire. The unreal came to rule the real. That was the first obvious and enduring achievement of the city.[3]

God, in the Judaeo-Christian tradition, wasn't impressed by the idea of conglomeration. The Tower of Babel – a quintessentially Mesopotamian structure – reached vauntingly towards God's own province, the heavens. God wasn't having it. He destroyed the structure. He deconglomerated. He splintered the consolidated people and their consolidated language – a sort of Mesopotamian Esperanto – creating many groups, and so many edges between the groups. It was a return to the status quo – to the small clan structures that had characterized pre-Neolithic life.

The deconsolidation didn't last.

Dick Whittington was a fool. The poor boy thought the streets of London were paved with gold. We're all Dick Whittingtons, more or less. Even if we live in a bothy looking over to Scotland's Outer Isles, and listen to the deer roaring in the glen rather than the juggernauts roaring on the ring road, and so are immeasurably rich, part of us thinks that the metropolis is where life really happens.

Whittington became Lord Mayor of London, and his journey is sometimes cited as a vindication of the metropolis. Not so: I'm sure he was a bitterly disappointed man. He was if he was honest. Just one golden paving stone would have outweighed a thousand times the gold in his chain of office.

The seductive centre always lies. About everything. About its motives, its finances, its potential, its influence and its record. It has never, ever, produced anything significant. It can't as a matter of principle. The centre sucks in. It wants more of itself. For the centre, bigger is better. What's the only economic mantra we hear? *Growth.*

Growth. Growth. The only uninhibited growth in nature is cancer. Cancer is our economic model.

It's a good thing to be small because it makes you edgier, and so more relational, and so in a more vibrant conversation with what's outside you. Conversations produce results. The same goes for nations: Leopold Kohr has demonstrated that the most creative and innovative nations in history have been small[4] (and so far as international order is concerned, we'd be all far better off with lots of tiny states – none too big to cause international problems, and each small enough for robust internal accountability).

Of course, as a matter of mere *geography*, novelty often comes out of centres. We shouldn't pretend that the Athenian Golden Age was not Athenian, that there was not much lasting poetry and resonant philosophy written within the walls of Rome, that Byzantium wasn't gilded or that Sancta Sophia has any peer, that Renaissance Florence was a backwater, that Venice just made mud pies, that Impressionism didn't flourish or Surrealism ignite in Paris, or that in the arts London's on a par with Des Moines or Oxford with Milton Keynes. But those examples, and others urged on us by the centrists to establish their credentials, are worth a closer look.[5]

Periclean Athens looks like the centre of centres. It had arrogated to itself much of the power and money of the Delian League, and had become the centre of gravity of the Mediterranean. Its playwrights were the funniest and freest anywhere until Shakespeare; its poets the most pungent and plangent; its politics the most enlightened; and its sculptors could turn marble to flesh. Yet, as Rex Warner says, its citizens, 'however careful they may be, live in a state of very great insecurity. They are poised between a number of tremendous and conflicting forces ... '[6] Warner gives the story of Hippolytus as the classic example. Hippolytus, the hunter, worships Artemis the huntress rather than Aphrodite. Aphrodite, affronted, makes Hippolytus' stepmother, Phaedra, fall in love with him. He resists Phaedra, but with fatal consequences.

Many of us feel that the gods treat us as Aphrodite treated Hippolytus. Or that they may do tomorrow. Cancers, auditors and HR managers are as capricious as the Olympians. But Athenian insecurity was even worse than ours. Those proud Athenian burghers, apparently so self-possessed and effortlessly able, were constantly looking over their shoulders, wondering when Sparta would march in from the west or Persia from the east.

They were looking, too, at the vines and olive groves on their farms. They weren't truly metropolitan people, but farmers with a pied à terre in central Athens, a helmet and a spear at the bottom of the bed in case they were needed in the night, and a whole host of ruling neuroses. Euripides' voice quavers. There's not a swagger anywhere in Aeschylus. The gods were in charge, and needed the blood of pigeons to dissuade them from drinking the blood of men. That most archetypal of centres was riven by fault lines running through souls and sleep, even if the body politic was fairly solid. Everyone was on a fault line, and knew it.

Rome, at the peak of its cultural prowess, was a mere collector of provincial boys and provincial ideas. The boys and the ideas mixed; the cocktail was extraordinary. But the provincials remained provincials. They brought their edges to the city, and the edges endured, sometimes altering their width and course slightly, but remaining edges. On those edges, giving the view and the insecurity necessary for lasting art and thought, perched the provincials. The only Roman thing about them was their postcode.

To be a collector is something, but it is not the same as being a parent. Rome itself hatched few distinctively Roman Romans. There are no major exceptions other than Cicero (and even he was born outside Rome). He makes the point well. That double chin; those wobbling jowls. The flatulent speeches. The prostitution of philosophy for political machination. The palpable enjoyment of intrigue for intrigue's sake. The cynical pragmatism. He is the eternal, universal occupant of the smoky room, to be found in a nicely cut toga or lounge suit wherever there is a mining concession to be bought. Yet

40

the forgiving edges redeemed even Cicero in the end. Brought to bay by the hounds of the Second Triumvirate, he uncovered his neck for the decapitating knife, ending with felicity an unheroic life.

Byzantium's own centre was, despite its ostentatious wealth and the murderous skullduggery of its courts, not only beyond its own walls but also beyond this world. In the marketplace, court intrigues were theologized, angels anatomized and Neoplatonic syllogisms perfected. Its shimmering mosaics invoked the transcendent, and whatever the transcendent is, it is over the quotidian edge. Its academies may have been truly metropolitan (though they were filled with dust-stained travellers from the Eastern Empire), but their syllabus was from the realm of the Platonic Forms.

Renaissance Florence might seem to confound this thesis. For there, in one place, in a fairly short span of time, was an unparalleled efflorescence of loveliness and iconoclastic power fuelled by the gold of typical centre people. It will not do to say, as I did of Byzantium, that this was not a centre-movement simply because it took a religious form, looking out to the stars and down to the catacombs. For though Renaissance art was overwhelmingly religious (one has to strain as herniatingly hard as Jacob Burckhardt did to see it as an expression of philosophical humanism), it did involve a rediscovery of the *human*. But, as the art historian Andrew Graham-Dixon has compellingly demonstrated, this rediscovery, which sparked the Renaissance, came from that most classic of edge-people, St Francis of Assisi: the little friar in the patched cassock who was flogged as he was dragged around town, talked with birds and made friends of man-eating wolves, mocked the rich, and whose identification with the ultimate edge-man, Jesus, was so complete that his hands, feet and side were miraculously pierced as he prayed out in the wild. He was the least likely ally of the Medici. Yet, long after his death, they bankrolled his agenda. They turned out to be his tools, not he theirs.

Francis, and later the Franciscan brotherhood, searched for ways to make Christianity accessible to the unlettered hoi polloi.[7] Francis began by announcing the compliment that Christianity paid to humans: Christ had divine dignity; and so too, whatever the factory foreman said, and even if leprosy had snapped off every finger, did each human. Francis's genius was to locate Christ's solidarity with ordinary people in the human body. Christ had a body; so did the people. Christ was born to a human mother, as were all humans. Francis invented the nativity crib to remind worshippers of Christ's vulnerability and accessibility. The transcendent Christ of Byzantium has a half-smile even on the cross. Francis made him bleed and scream, just as the people bled and screamed when they died, and from Giotto onwards, the artists and sculptors of Renaissance Italy made him bleed too.

The Franciscans got their great evangelistic opportunity when an industrial revolution, centred mainly on textiles, caused an expansion of cities in Tuscany and Umbria. Workers flooded to the edges of the cities, and the Franciscans ministered to them there. The Franciscans built churches fast – out of brick rather than stone – and invented the art of fresco (frescoes could be completed in weeks rather than the months or years needed for traditional mosaics) to dramatize the story of salvation. Look at a map of the original Franciscan churches of Florence – they are all in the disreputable edge-lands of the city.

Renaissance art's rediscovery of the body was a direct response to the market created by the Franciscans. The Renaissance revolution was not a declaration that Man was the measure of all things, but rather that Christ was the measure of Man. No one had depicted realistic human bodies since the classical period, and artists had forgotten how it was done. To remind themselves, they literally disinterred the human form from the soil of Italy. The collection of Roman sarcophagi at Pisa was a major source: Roman faces and limbs were copied from there onto Nicola Pisano's great pulpit in the Pisa Baptistery (1260). From there human body parts spread across Italy, declaring the mystery and vaulting status of incarnation. The realism was an overt form of Christian polemic. It worked. Even the most down and out could look at Donatello's heartbreaking *Penitent Magdalene*, which shows the Magdalene as a tangle-haired bag-lady, and say: 'If *she's* a saint, I can be one too.' The churches bulged.

Only Venice escaped the Franciscan invasion. It has always been more of a Byzantine than a Western city, with the evanescent light on the lagoon reflected in its mosaics and its appetite for the transcendent. Its crucified Christ carried on smiling. But, as Graham-Dixon suggests, perhaps there is a more prosaic explanation for its artistic isolation from the great Tuscan currents. Its geography has always deterred invaders, and its humidity frustrated the invasion of Renaissance realism. Its walls are just too damp for frescoes, and so it would have had to hang on to its mosaics even if it hadn't been ideologically attached to them.

But what about the sponsors? Those canny bankers who paid for Leonardo and Michelangelo and Raphael; without whose money Jesus might have kept his half-smile and his place in the high heavens rather than being summoned down to the slums of Florence. They paid his fare from Paradise. Don't they make a nonsense of the thesis?

We see the Medici as fat spiders in the centre of the Tuscan web. They were. But they were terrified of damnation. (At least until Alessandro de' Medici, who doesn't seem to have been frightened of anything in heaven or on earth.) They read their Bibles carefully, and knew that the prognosis for usurers was not good. They sought to buy off the risk of hellfire. Yes, there are some vainglorious palaces in Florence and its environs, but the jewels in the crown of the Florentine Renaissance were bought with bankers' soul-money. The piety of the Medici looks hypocritical to us, but it was real. Cosimo de' Medici paid for the Dominican monastery of San Marco. In every cell there is a fresco by Fra Angelico. Cosimo had his own cell – a rather luxurious two-storey affair. Above the door is an inscription recording the Pope's promise to him that, in return for founding the monastery, all his sins would be forgiven. The Medici were prompted to write their cheques by their timorous look over the edge of this life and into the next.

It wasn't only the bankers who were fearful; all late

fifteenth-century Italy was consumed by millennialist terror. The Earth was poised, they thought, on the edge of time itself. That was bound to concentrate attention and generate energy, inspiration and sponsorship.

Under Lorenzo de' Medici, Florentine art (particularly under the chisel of Michelangelo) took a more secular turn[8] – but it was only *relatively* secular. It was still a down payment for salvation. It's significant that when the impious, sacrilegious Alessandro assumed the Medici throne, the quality of the art plummets. It starts to clunk.

Even the undeniably great artists moved around a good deal: Leonardo from the hamlet of Anchiano to Florence to Milan to Rome to France; Michelangelo from Caprese to Florence to Venice to Bologna to Rome; Raphael from Urbino to Florence, around northern Italy, to Rome, and so on. We tend to think of Florence as the epicentre, and in many ways it was (thanks to the little dead friar and his banker-pawns), but it had no monopoly on genius – whatever that is. It's the same for the other Renaissance centres.

Paris? Is there such a place? 'It' is and was a dense archipelago of salons. Its artistic schools owe nothing to the city, if it exists at all. They sought to make universal statements, untethered to anywhere on the planet, about light (in the case of the Impressionists) and about epistemology and the nature of reality (in the case of the Surrealists). Their genius might be characterized as a repudiation of the place they worked. It was certainly not an endorsement. The most influential Impressionists lived not in Paris but by a little lily pond in a back garden in Giverny, or in a house in Tahiti, watching the light on the Pacific, or in a madhouse in Saint-Rémy-de-Provence. Their most famous exhibition was organized because they weren't selected to participate in the 'official' Salon.

And London? Here I will be accused of being selective; of skewing my sample. But I sent an email to lots of people, asking: *Who are the distinctively London artists? By which I mean those artists, in any medium, whose work could not have taken anything like the form it did if*

44

the artist had not been from London. The answers were amazingly consistent. No visual artists were regularly mentioned (though Hogarth put in an appearance), but five writers' names kept recurring: Chaucer, Samuel Johnson, Blake, Dickens and Chesterton.

I can take them shortly. Chaucer's fame rests on *The Canterbury Tales* and *Troilus and Criseyde.* Neither smells distinctively of London – *Troilus* very obviously not. London is necessary to the *Tales*, but it is necessary as a place to *leave*, not a place to be. His pilgrims are from all over the kingdom, and they are going to an obscure marshy town in Kent which just happens (more of this later) to be the heart of English Christendom. They are released by the road, released by the departure, released by being out in the sticks, to be more themselves than any other characters until Shakespeare (another dowdy provincial). What's more, Chaucer's style – though misrepresented as canonically English – is far more French. He imported iambic pentameter, repudiating the rollicking, rolling alliterative verse of the provincial *Gawain* poet.

Johnson? He sits there in his Fleet Street garret, depressively coining epigrams and fussing about words. Might he not have done the same, or the same but rather better and less miserably, up a tower in the Dordogne like Montaigne?

And Blake, who lived almost his whole life in London? He, after all, believed that 'Only in London' could he do his writing and his art. But where did he really live? Not in a place recognizable as London by any of the centre people. There were angels in every corner; spirits hunched on every shoulder. His conversations were not with shopgirls but with the Archangel Gabriel. 'The fields from Islington to Marylebone to Primrose Hill and St John's Wood were builded over with pillars of gold,' he wrote, 'and there Jerusalem's pillars stood.'[9]

As for Blake, so for Chesterton. He was built of the cakes of Blackfriars and the ale of Fleet Street. But his London – his whole cosmos – is, like the city and cosmos of Byzantium, infused with light from another place, and only by that light can Blackfriars and Fleet Street be properly seen.

Dickens is the London chronicler par excellence. I do not say that he is not a distinctively London writer (though much of his life was spent elsewhere), or that his London books (far from all his books, of course) could have been set anywhere else. His writing, more than that of anyone else, has Thames water running through it. But it is by being *the* distinctively London writer that he makes my point. For he wastes no words lauding metropolitan institutions. Whenever he sees a centre he shouts 'Fraud' or 'Confusion'. At the very centre of London, at the centre of the majestic law of England, is obfuscating, debilitating, pea-soup fog:

The raw afternoon is rawest, and the dense fog is densest, and the muddy streets are muddiest near that leaden-headed old obstruction, appropriate ornament for the threshold of a leaden-headed old corporation, Temple Bar. And hard by Temple Bar, in Lincoln's Inn Hall, at the very heart of the fog, sits the Lord High Chancellor in his High Court of Chancery.

Never can there come fog too thick, never can there come mud and mire too deep, to assort with the groping and floundering condition which this High Court of Chancery, most pestilent of hoary sinners, holds this day in the sight of heaven and earth.

On such an afternoon, if ever, the Lord High Chancellor ought to be sitting here, as here he is, with a foggy glory round his head, softly fenced in with crimson cloth and curtains, addressed by a large advocate with great whiskers, a little voice, and an interminable brief, and outwardly directing his contemplation to the lantern in the roof, where he can see nothing but fog.[10]

It is not a great affirmation of the Establishment.

Dickens' interest is in fringes: in orphanages, brothels and sweat shops; over the edges of convention, propriety and endurance.

In the great urban centres there is indeed innovation and

achievement. There are new syntheses. But they come always from edge-people who remain edgy and sharpen themselves by grinding on other edges. Edges criss-cross cities and may constitute meshes of edges in which it is possible to live, thrive and create.

Throbbing, cosmopolitan metropolises are, at their best, constellations of edginess. 'For me,' says the Australian poet John Kinsella, 'all centres are fringes.' Tony Judt saluted edges, and in the same breath saluted great cities. He mourned modern urban changes because they had become more uniform – less edgy. 'I prefer the edge,' he wrote:

> The place where countries, communities, allegiances, affinities, and roots bump uncomfortably up against one another – where cosmopolitanism is not so much an identity as the normal condition of life. Such places once abounded. Well into the twentieth century there were many cities comprising multiple communities and languages – often mutually antagonistic, occasionally clashing, but somehow coexisting. Sarajevo was one, Alexandria another. Tangiers, Salonica, Odessa, Beirut, and Istanbul all qualified – as did smaller towns like Chernovitz and Uzhhorod. By the standards of American conformism, New York resembles aspects of these lost cosmopolitan cities: that is why I live here.[11]

Geographical centres are magnets – they attract edge-entities like iron filings. They are chatrooms where edges meet. They are petri dishes where edge-organisms multiply. They are incubators where neonatal edges are nurtured. They are debating chambers for the ventilation of ideas. They contribute little to the debate. A tired painter in London, out on his afternoon stroll by the Thames, might go as far as Westminster, smell the decay and, convinced again of the importance of his work, run back to his nihilist easel in a Hoxton warehouse.

This, at least, is my experience in the centre I know best: Oxford.

If my thesis doesn't hold, then writing this book is appallingly hypocritical. I am a Fellow of an Oxford college, right in the heart of Oxford. It was founded in 1314, over a century before Agincourt. We eat in a high-timbered hall, and drink claret in an oak-panelled room hung with seventeenth-century portraits. The choir sings Tudor anthems at Evensong in the chapel. It is the happiest, most functional and most fecund institution I've known. To anyone looking in from the outside it must seem smug and comfortable – the apogee of the centre.

Half a mile from the college gate, the river Thames flows to the steps of the House of Commons and then on to the City of London. Far too many of the students follow it and join the institutions of the centre. They forget the lessons in scepticism they learned here, and go off to become part of the problem rather than the solution. But I don't blame the college for that. It does its best to inoculate them against the infection of the centre: to teach them that there is no account of human thriving ever articulated that is consistent with working for their corporations.

The college itself is exciting and productive precisely because it is populated by fringe people. I've never met anyone there who isn't crippled, as I am, by impostor syndrome. I wouldn't dream of eating with anyone who didn't share my own (grave) diagnosis. At dinner the clinking of spoons in soup bowls is drowned by the thunder of invigorating insecurity. That's what you need for scholarship and proper fellowship. Unless we know we're in the same sinking ship, can we really have a meaningful conversation?

If, by a bad 'centre', I don't mean a geographical centre, or a cultural centre, or a metropolis, what do I mean? I mean a hive of the conceit that comes from believing that one is the centre oneself. This conceit tends to inhabit geographical or cultural centres. It is characterized by a sense of entitlement and unaccountability. In the realm of statistics, as we will see in Chapter 11, it tyrannizes by averaging. Where it parasitizes a human soul, it is obliged to nothing and no one. It lives in and is sustained by membership of cabals – but cabals not of friends, for the sacrificial reciprocity that is the prerequisite of friendship is impossible. Its sound is braying. It has no real words, because it has no convictions. Its theology is pragmatism. It is immune to true criticism, because to be criticized one has to believe in something. It is closely identified with the Establishment as that term has come to be understood over the last couple of decades. (That Establishment of course is associated with the place where my own soup spoon nervously clinks. I can only say that the existence and nature of the Establishment represent a shameful failure of Oxford: a failure to teach the humility and uncertainty which are the start and endpoint of all proper scholarship.)

This Establishment, even when it is notionally political, is not truly so. Politics, after all, implies a political philosophy, and philosophy still hasn't completely forgotten its etymological roots – 'love of wisdom'. In the centre there is neither love nor wisdom. The centre's love of Self generates its main desire, which is control for control's sake and influence for influence's sake. The best way of achieving that is incontinent self-multiplication. The centre implants its memes into every brain and every journal, insinuates its acolytes onto every board, and puts its face on every page. It hates and fears true variegation, because that would mean that most of the world was populated by things other than itself.

When it is crossed, the centre cuts up rough. We will see later just how rough.

It was Sarah's last night as a student. She had just completed her doctorate on a fourth-century Syrian ascetic noted for his austerity of life, sweetness of character, and his savage dismissal of the demons who tried to persuade him to give it all up and head back to Damascus.

We sat by the river, drank bad wine, ate curling sandwiches salvaged from someone else's picnic and talked about the future. That was all she wanted to talk about. She was excited. In a week's time she was starting, with a mind-boggling salary, at a firm of management consultants in London. 'You've got to go there really, don't you,' she said. 'It's where everything happens.'

It was a throwaway remark, but revealing in its casualness. Everyone seemed to make that remark sooner or later, but Sarah hadn't ever done before what anyone else did. I enthusiastically agreed and followed her to London soon afterwards, joining everyone else in the refrain. It was years before I began to examine those words: *'It's where everything happens.'*

My interest here isn't primarily in the consequences of that sort of thinking – solemn though they are. Some of the consequences are well documented, much discussed and politically explosive: underinvestment in the provinces, the difficulty of selling the 'levelling-up' agenda to the centre, and so on. I'm interested in what the observation says about our view of the centre, the centre's view of itself, and the centre's view of the non-centre.

Sarah was – and is – no fool or charlatan. She had – and still has – a healthy view of the chicanery of the political establishment and its acolytes. Yet she thought the train to London as inevitable as ageing and death – and about as fundamental to human life. You just had to deal with it somehow. There was no point in protesting. You might as well try to move a mountain.

She summarized accurately the centre's view of itself and of the world outside the centre. 'It's where everything happens' is an explicit denigration of most of the lives in the United Kingdom – and probably, since the United Kingdom is of course the centre of the universe, of all the non-London lives anywhere. If 'everything' happens in London, nothing happens anywhere else: nothing, anyway, worth noticing, let alone investing in.

Clumsy statements like Sarah's are morally toxic because they tend to conflate events with moral significance. They say that we are what we do, and if we don't do anything that's regarded as anything by the centre, then we are nothing ourselves. The implication is that if the provincial town hall doesn't host West End productions, provincial lives aren't as valuable. Once seeded, that idea yields a great crop of ethical and political tares. For if the lives aren't as valuable, it's not wrong to fail to enrich them. West End shows would be wasted on them, and it's fine to maintain the status quo of metropolitan hegemony.

There's another poisonous idea lurking in the clumsiness: an idea about causation. Sarah wasn't just saying that London was where 'everything' happened, but where 'everything' was *made* to happen. She had bought into the centre's belief in its own omnipotence. If only the centre has agency, nothing else does, and it's not culpable to ignore (or worse) the non-centre. It's a notion of causation allied to the distasteful 'Great Man' theory of history: a notion enthusiastically bolstered by those in our modern centres who consider themselves Great Men.

Two theories of history explain most things. Neither has any truck with the posturings of the centre.

The first is the Great-But-Unacknowledged-Women Theory of History. This has two variants. One is the Power-Behind-the-Throne variant. Think of *My Big Fat Greek Wedding*. The man might be the head, says the mother, but the woman is the neck, and the neck can turn the head in any direction it wants. Since women are more holistic and relational thinkers than men (and so bring more information to bear on their decisions), the policies emerging from the men at the centre will be less bad than they otherwise would have been, because the women will have been the main authors. A variant of this is the Life-Doesn't-Happen-in-the-Centre-Anyway-And-Where-It-Does-Happen-Women-Are-The-Main-Ones-Who-Make-It-Happen theory. We'll see an illustration of this in a moment in a brief visit to thirteenth-century BCE Egypt.

The second major theory of history is Nassim Taleb's notion of the black swan – we don't expect swans to be black – which holds that, by and large, history doesn't crawl but jumps, and the jumps are wholly unpredicted and unpredictable.[12] Sir Francis Bacon (1561–1626) got it right: history's most important advances are the unforeseen ones, those 'lying out of the path of the imagination'[13] The past is a dismal predictor of the future, yet we cling to the belief that things are going to happen as they happened before, and so, by planning (the forte of the centre and the pride of the centrist ego), we can direct the future.

We can't. We're tied to antediluvian notions of causation, always responding to old happenings. The French built the Maginot Line based on the previous German invasion route. Hitler simply went round it. After the stock market crash of 1987, half the traders in the US braced themselves each October for another crash. That was foolish: there was no precedent for the first one. I have friends in the City (yes, I still have friends in the City, though perhaps they won't survive the publication of this book) who still can't sleep in October. But the next black swan will be unimaginable. Taleb uses the example of 9/11 – an example shocking in its simplicity. 9/11 happened precisely because it wasn't supposed to happen.

One corollary of this is politically important. It means, as Taleb points out, that reading newspapers and watching news programmes *decreases* our knowledge of the world. (It may increase the amount of *information* we possess, but that is a *very* different matter.) Apart from the cultural bits and pieces, there are two main types of material in newspapers: old news, and speculation based on old news and on ex post facto explanations of past news. Both, as far as knowledge of today's and tomorrow's world is concerned, are almost worthless. This matters because of the alliances between centrist political power and some elements of the news media (we all know who I mean). Those alliances are rightly criticized because of the spin the media put on the events. But there is a more fundamental reason for concern: the media's historiography is ludicrous.

The social sciences have largely adopted the fallacy of thinking that events are predictable. It's not surprising. Indeed, it is inherent in the very name of the discipline: social *sciences*. This implies that there is a way of measuring and accounting for the uncertainties of society and therefore history. There isn't. The social sciences should be renamed to reflect the real nature of their subjects. They purport to study humans, and so I suggest the name 'Humanities.'

We have seen already, and will see again, some of the problems of the centre's beloved top-down theorizing and legislating. For now I'll simply endorse Taleb's observation that top-down planning demonstrably doesn't work. Karl Marx and Adam Smith were both wrong. Free markets work, when they do, in the same way that evolution works: by giving free rein to trial and error, by allowing people to respond to the landing of a black swan on the trading floor. There is a limited role for social *science:* to describe (and perhaps to measure – which must be done retrospectively) responses to black swans. But that sounds rather like the discipline of history. If it includes prognostication, with or without the help of a statistical model, it's not science, nor is it any other *discipline*, but fortune-telling, and should be banished from university departments to fairground caravans.

Yet this very prognostication is at the heart of the claims of centrist governments to govern. It is the bottom line of all speeches of would-be Great Men of history – of whatever sex – when they tell us why we should send them to the centre to rule.

I've said enough about our modern contexts, and perhaps, as my daughter Rachel (whom we'll meet shortly) found in Parliament Square, we're too close to them to see them clearly. To illustrate that what I'm saying about centres is *general*, arising from the nature of humans and the fabric of the world, let's go back four thousand years to a context unfamiliar to most of us. To Egypt. That should give us enough distance for a proper view.

There's a lazy tendency to assume that to tell the story of kings and queens and governments is to say something meaningful. It's not. Life happens despite monarchs and viziers, not because of them. And it happens a long way from them. Most children aren't born, most sausage and mash isn't cooked, most poems aren't written and most people don't die in Cabinet rooms; and what goes on in Cabinet rooms has nothing whatever to do with anything other than the price of being born, the quality (in some very limited respects) of lives outside the rooms, the price of being buried, and the price and quality of the sausage and mash. Revolutions don't start in parliaments, but in bedrooms, kitchens, and in that most marginal of all places, the human heart.

A good, if dispiriting, example of our tendency to equate rulers with history is in the standard chronology of ancient Egypt. There are the three great epochs: the Old, Middle and New Kingdoms. The often explicit assumption is that these are the times when anything and everything that was important happened. Kings and queens are the ultimate centrist figures. The Egyptian monarchy lived and ruled in one or both of the great centres – Memphis in the north, near modern-day Cairo, and Thebes in the south (modern Luxor). In this chronological scheme, the monarchy is equated with the country. The life of a peasant in the Delta, five hundred miles from Thebes, is assumed to be merely a distant part of the Theban court. It's true, of course, that there was a strong monarchist theology which saw the stability and prosperity of Egypt as tied up with the monarchy. Perhaps some of this theology seeped into the peasant's hut; perhaps, at the time of a strong monarchy, it gave some comfort. But I doubt it was the well-spring of his being; the ground of her ontological confidence. And whatever the bias in the history books, there were always far, far more peasants than kings, queens, priests or viziers.

Sandwiched between the Kingdoms are the Intermediate Periods. The language is illuminating. It's saying that these periods are not *real* Egyptian history. The country was just marking time until it got back to the real royal business.

What really happened in (for instance) the First Intermediate Period (2181–2055 BCE)? There was decentralization. The provinces became more important and the status of the traditional centres diminished. That's why it's not regarded as Egyptian history proper. The narrative is massaged by quisling historians to suggest that a lack of central control was (and generally is) disastrous. It's usually said that the Period was a dark age between two golden ones, and was characterized by war, social chaos, economic decline and wretchedness.

Of course it suited the subsequent monarchies to paint the Intermediate Period this way. By and large, the Middle Kingdom's own propaganda has been uncritically rehashed. That propaganda says (literally) that the sun wasn't as bright in the Intermediate Period; that the Nile gave up on Egypt and didn't inundate as regularly or as generously as it did when an anointed king was on the throne. But that's spin. There were undoubtedly difficulties during this Intermediate Period, but the surviving texts from that time which detail the crises are upbeat – they speak of the crises being *conquered*. They are vigorous, proud and self-confident.

Here, for instance, is what Ankhtifi, a ruler of a province in Upper Egypt, said about himself:

Horus brought me to the Horus-Throne nome [Edfu] for life, prosperity, health, to re-establish it, and I did. For Horus wished to re-establish it because he brought me to re-establish it. I found the House of Khuu inundated like marshland, abandoned by him who belonged to it, in the grip of a rebel, under the control of a wretch ... I am the van of men, the rear of men, for my like has not been, will not be, my like was not born, will not be born. I have surpassed the deeds of my forebears, and my successors will not reach me in anything I have done for the next million years.[14]

We should be as suspicious of these First Intermediate accounts of itself as we should be of the Middle Kingdom's self-serving texts. But there are very good, and increasingly acknowledged, archaeological reasons to prefer the Intermediate Period's version. The Egyptologist Stephan Seidlmayer summarizes the emerging consensus this way:

The periods of war and conflict that make for such startling reading in biographical narrative were therefore no doubt only localized and short-lived episodes. At most places, for most of the time and for most of the people, the First Intermediate Period probably would have been a rather less thrilling experience.[15]

Most people just got on with their lives. They didn't need the centre to do that. And *we* don't.

But life is more than eating, drinking and copulating undisturbed. Reflection, art, innovation and 'high' culture contribute to human thriving. Don't you need vigorous central patronage for that? The conventional historians, at least implicitly, have tended until recently (there's change afoot) to assume that you do.[16]

In the case of the First Intermediate Period, it is clear that, if anything, the absence of central control was a splendid thing for culture. Seidlmayer again:

> Far from being a period of cultural decline, these turbulent years witnessed an upsurge of outstanding creativity, adapting and developing the existing media of literary and pictorial expression to correspond to a new range of social experiences.[17]

The people thrived on their edges. We shouldn't be surprised.

It's my generation, and the one before it, that remain most in thrall to the power of the centre. For all my moaning about my teenagers, they see attraction to the centre as an obstacle to trust and a disqualification for power. The perceived apathy of Gen Z isn't apathy at all: it's a passionate, clear-sighted rejection of the centre. It comes from seeing how things are – and not just politically. Politics are just an illustration of the general principle.

There are three splendid examples of modern centres. First: establishments in the bad sense – the pottage of political power and economic influence for which so many sell their birthright, and which is served up in boardrooms and the committee rooms of the House of Commons. Nothing new comes out of these places. If you think you've seen something new, look more closely. At best it will be an emasculated iteration of something dreamt up by edge-people. The centre can't create; it can only twist and corrode. That's true not just of its productions in terms of policy and ideas, but also of its human inhabitants. The centre's a perilous place to live, because its ultimate appeal is to the Self.

And that's the second example. About the Self I need say nothing, because we all have one, and we all know, if we're sufficiently reflective, what self-centredness brings, and if you don't, nothing I can say will make any difference.

The final example is the black hole. Black holes are the apotheosis of centredness; they are so heavy that they affect the orbits of everything for millions of miles around. Anything drawn into them is crushed and ceases to be. Their pull is so great that not even light can escape.

5

The Machine State

Such a shame! Though destined
for great things, your wretched destiny
always unfairly betrayed you,
denying you both cheer and success,
baffled by commonplace trivialities,
by pettifoggery and by indifference.
An evil day it is when you succumb
(the day you let yourself succumb)
and take the Royal Road to Susa
and offer fealty to King Artaxerxes
who kindly grants you courtly sinecures,
proffering satrapies and other suchlike gifts,
which you accept in desperation –
these baubles which mean nothing to you.
Your soul hungers and craves for other things:
for the acclaim of City and Sophists,
that priceless, hard-fought for adulation,
for the Agora, Theatre and Laurel Wreaths.
How can an Artaxerxes give these to you,
where will you find them in your satrapy
and what will life be worth without them?

CONSTANTINE CAVAFY,
'The Satrapy', trans. John Stathatos[1]

I MET JIMMY ON A road somewhere near the border between Thailand and Myanmar. I was riding a motor scooter. He stepped

out, held up his hand to stop me, and asked in perfect English for a cigarette. I didn't have one.

'Ah,' he said. 'I didn't think so. But you have hurt your leg.'

I had. I'd come off the scooter a few miles back and was bleeding. It wasn't serious for me, but it was for Jimmy. He disappeared into the undergrowth at the roadside and emerged with a handful of leaves. He mopped my grazes with them, stood back to admire his work, and said, 'Now you will live.'

I'm not completely stupid. I know a tourist-exploiting tout when I see one, and there was one here.

He could have been anything between twenty and ninety. He wore ragged khaki shorts, flip flops, and a T-shirt declaring *Coke Is It*. He was bandy-legged, as if he'd been born on a horse, and had mocking gargoyle eyes and jug-handle ears. His mouth hung open, because it was easier that way.

'So,' he said. 'You want a guide.'

It wasn't a question. I didn't, but because this was his patch, not mine, and it seemed rude to contradict him, we agreed a price, which was – I knew it even at the time – about five times the going rate. For this I would have his company, his wisdom, his cooking and the loan of his uncle's nearby shed to store the scooter until I came back.

'So, what do you want to do?' he asked, after everything was agreed. You might have thought this should have been the first thing discussed. It never occurred to either of us.

See something of the hill tribes, I told him. And none of the ones on the regular trek routes, thank you, where couples get married every night so the tourists can experience the warmth and colour of traditional village life. 'Fine,' he said, and after tea in the uncle's shed we climbed on a bus that took us seven hours somewhere.

Jimmy was true to his word. In the fortnight we spent together there were no fake weddings, no histrionic masked dances with cute children sent round with a tin to beg *baksheesh*, no salubrious bunkhouses with Wi-Fi and flush toilets. Just birds, barking deer curry, the whirr of cicadas and the hum of the planes sent to strafe the opium poppies. But I had forgotten most of that until I went back to look at my notebooks. What I remember is Jimmy.

'Is your name really Jimmy?'

'It's as good as any other, isn't it? And better than some.'

'Where are we?'

'We're here.'

'Where's "here"? What's this village called?'

'No idea.'

'How long has it been here?'

'Who knows? A week? Ten thousand years? Probably won't be here tomorrow.'

'Are we in Thailand?'

'Could be.'

'Or Myanmar?'

'Could be.'

'Well, do the people think they're Thai or Burmese?'

'Why should they care? Why do you care?'

'Well, it has consequences.'

'Really?'

'Where do you usually live?'

'Around.'

'Are you married?'

'Sometimes.'

And so on.

He delivered me back to the uncle's shed, counted out carefully – and twice – the money I gave him, bowed and evaporated.

I must write about nations and politics. I don't want to.

There are three reasons. The first is simple distaste. I feel dirty if I read any political news. Howard Jacobson writes about the echoing sense of emptiness he gets when he talks about politics. I'm the same, but worse. The emptiness makes me nauseated, as bad hunger can do. Almost everything important and sustaining is necessarily missing from a political conversation. I take it as self-evident that most politicians are self-serving charlatans. There are just enough exceptions to show that being a self-serving charlatan is not inevitable, and so prove the culpability of the majority.

That brings me to the second reason: states don't really exist. They are fictions; fables; vanities. I don't like being bossed around by non-existent entities. At most they are ephemeral, and it is a waste of precious time paying them any attention.

The third reason is that politics don't – and can't – address anything that really matters. We've seen something of this already, on our visits to ancient Egypt and the UK Cabinet. Unless I let them, politics can't dictate – though they may try – what I think or feel. They can't stop me loving the good or hating the bad. They can kill, lie and enslave, and though those things have real consequences, which of course I don't belittle, they can't make or unmake me. They can make me hurt, curse and lament, but though they might kill me, they can' t – or needn' t – dehumanize me.

Politics can't really hurt me because they can't see me. I – I as I really am – am invisible. You can't see from the centre anything as elusive and distant as a human soul. Politics are all about theory and power. No political theory ever devised can say anything accurate about a creature as enigmatic as a human. No human power can do anything to shift anything as vast and weighty as a human if the human chooses not to be moved.

I'm alarmed to find myself agreeing with Roger Scruton, who, when explaining why he thought that most intellectuals are left wing, suggested that part of it was due to an over-great love and trust of theory. Genuinely appalled at the wrongness of the world, they see the exercise of power as the remedy. Power is prescribed in accordance with some theory (Foucauldian or whatever), and individuals are subsumed into the theory. They are inconvenient clutter. They can and should be ignored. The real thing's the theory. Political theory of this kind is a dance of paradigms. No true 'I–Thou' encounter is possible in this dance: there are no erotic or interrogating glances as in a real quadrille, as the partners wheel around one another, shoulder-tip to shoulder-tip. And 'I–Thou' is what the cosmos is all about. Power misses the whole point.

The sort of power that's found in the centre, anyway. There are abuses of power outside the political metropolis, of course. In the yurt where the decisions about that season's route are made. In the family kitchen. In the bedroom. But they are not backed by the august authority of a national parliament or the forbidding footnotes of critical theory.

It might sound as if I'm advocating a conventional small-or-no-government neoliberalism, or an anarchy. Not so. Anarchy didn't build Chartres Cathedral, write *King Lear* or draft the European Convention on Human Rights. I'm with P. J. O'Rourke:

> . . . every bull-dorm session anarchist should spend an hour in [1980s] Beirut. So-called Western Civilization, as practised in half of Europe, some of Asia and a few parts of North America, is better than anything else available . . . Our civilization is the first in history to show even the slightest concern for average, undistinguished, none-too-commendable people like us. We are fools when we fail to defend civilization, [of which government can and should be a part] . . . [2]

I'd add: every neatly besuited neoliberal should spend a week in a cobalt mine in DRC. No: government has its place. It's probably better if that place is geographically nearer the areas it's supposed to affect, but my aversion to the centre isn't so severe that I'd want to insist that everything is devolved. The important thing is that the centre knows its place – which should be lowly. The role of government is to facilitate the thriving of people – not the thriving of an idea or policy. Bring us into the centre – whether it's into a boardroom, a parliament or a theory – and we stop thriving. The business of government is the meanest of all human occupations. It should be restricted to the provision and mending of actual and metaphorical roads that facilitate travel to the actual and metaphorical edges.

'Mr Wilcox glanced at Parliament contemptuously,' observed E. M. Forster. 'The more important ropes of life lay elsewhere. "Yes, they are talking again," said he . . .'[3]

But I can't leave it there. What you've just read – if you stayed with me through my sanctimonious, Western-middle-aged-male-wrapped-in-cotton-wool-and-living-in-a-nice-house-ishness – is far too easy to write. No amount of earnestness about the moral significance of humans makes it tolerable that there are slaves in the DRC toiling underground so that my children can scroll on Instagram or, worse, my not-really-elected representatives can plot on Google Meet.

If centrism is responsible for these ills, or if edginess can mitigate them, I need to overcome my distaste and get political.

There are two particularly powerful contemporary voices who have demonstrated the toxicity of the centrist state and proposed an antidote: the late anthropologist James C. Scott and the writer and poet Paul Kingsnorth.

First I must dispose of my tautology: *centrist state*. States by definition are centrist. Kingsnorth writes:

> The momentum of a state is always towards the centre; always towards the agglomeration of more power. No 'conspiracy theory' is necessary for any of this to be true, and neither do the people running the state need to be evil. It is simply the logic of the thing. A state is like a vortex or a black hole: at a certain point, it begins to suck in everything around it. As it grows, it will tell stories that justify its existence. Democracy, liberty and progress are some of the more recent banners that state power has gathered beneath, but there have been others: racial or ethnic homogeneity, human equality, religious purity. All of these stories have the potential to unite a people around a state core.[4]

This is not simply a matter of opinion; not the expression of a political conviction. It can be measured. We – wherever we are – are more surveyed, regulated, medicated, assessed, shepherded and controlled than we have ever been, and the centripetal force is increasing all the time. Rail against it and you're called a fascist or a communist. If it's during a global pandemic, you might be fined or told you can't walk in the park. Didn't the centre just love the pandemic?

Covid showed us what centres are really like. It showed us too what happens if you let the centre write the story that justifies its own existence. It will gladly pick up its pen. At the core of the problem of the modern state is the absence in our personal and societal lives of a story better than the one drafted by the state. We must look for a better story; one that is historically and biologically true, and makes us empathetic and humble.

It won't do to distinguish too neatly between the state and the devices (such as corporations and technologies) it uses to suck in power. They all share the same centripetal physics. Kingsnorth uses a single word to describe the coalition: the 'Machine.' It's an eloquent term. Whatever a machine is, it's not human. The Machine, as Kingsnorth understands it, is necessarily at war with everything that is most distinctive about humanity. To be human is to fight against the Machine.

There are different ways of fighting. We can't, thinks Kingsnorth, oppose the Machine head-on. It's just too powerful. And anyway, there's no need. The Machine has overstretched itself. Its appetite is unsustainable. It has started to eat its own children, and you can't live long on that diet. Kingsnorth, changing the metaphor, sees the state-corporate-techno creature as a great, lumbering, doomed dinosaur. We're the little furry mammals scurrying around its feet. We're too small to be seen, but we'll inherit the earth. The future's ours. We just need to wait for the meteor and try not to get trampled before it comes. Dinosaurs died because they couldn't cope with what lay beyond the edge of the catastrophe. But edges are our natural habitat. We're the edge-people. We'll be fine.

Our resistance should not be purely passive. Indeed, it cannot be. For if we merely live (*merely*! What a stupid word for creatures like us) as proper, free humans, we'll enrage the Machine all the more. It'll overheat, blow a gasket, eat more children, and die sooner. One of the most effective ways of upsetting and overheating the Machine is to become invisible. We've noted that humans are far too big and intricate to be truly seen by the dim eyes of political animals, but those eyes can see the marks representing humans that they make in their ledgers: the marks they make to count us, corral us and snoop on us. Let's hide the marks. If the Machine's spies and its infrared sensors suspect we're out there, but there are no entries in the spreadsheet, there'll be a very gratifying sound of cogs grinding, gears crashing, and we'll be freer than we were.

We can argue about when and where the first state was created. Most people would plump for Mesopotamia around six millennia ago. Nation states, as we know them now, are *very* recent institutions. But whenever and wherever it was, states of any kind are relatively new. For most of our history as a species, we have done – and done well – without them. States presuppose sedentism, or at least relative sedentism compared, that is, to our constitutional hunter–gatherer ways of life. Historically, most normal people have bitterly resented both sedentism and statehood. People tend to give up their wandering, edge-transgressing lifestyles only if they're desperate. When we can choose, we choose the margins.

Kingsnorth draws deeply on the work of James C. Scott, whose 2009 book *The Art of Not Being Governed* surveys the hill peoples of South-East Asia.[5] Over the last couple of millennia, the local states have tried, as states always do, to control their populations. But normal, natural, free people don't want to be controlled, and many fled to the hills where they could be administratively, fiscally and often actually invisible; where they could run their own lives. Freedom involves being sociologically and politically fleet of foot; being shapeshifters; doing whatever's necessary to dodge the state whenever it plods up the mountain pass, as long as that doesn't violate your defining story. If freedom involves becoming so complex that it creates an unbearable bureaucratic headache for the state, become complex. If it means evaporating into the tree-tops, where the bean-counters can't climb, evaporate. If it means becoming a bigger entity, become bigger; if smaller, then smaller. It involves using your freedom and valency to frustrate a sclerotic monolith.

Scott cites Lois Beck, who studied tribal culture in Iran, where the same tactics are used:

> Tribal groups expanded and contracted. Some tribal groups joined larger ones when, for example, the state attempted to restrict access to resources or a foreign power sent troops to attack them. Large tribal groups divided into smaller groups to be less visible to the state and escaped its reach . . . Such local systems adapted to and challenged, or distanced themselves from, the systems of those who sought to dominate them.[6]

In South-East Asia, Iran, and everywhere we care to look (including online communities and allotment associations and folk sessions in pubs and publishing houses that don't just churn out the edicts of the Machine), you see the same thing: shift, re-form, hide, sing your own song, see breathing as an act of rebellion, rebel not out of cussedness but because rebellion is the human condition and because the constant crossing of edges is what human feet were designed to do.

We needn't look to obviously 'alternative' cultures for examples of freedom flowering under the nose of the centre. Sometimes it flowers in the centre's name. Take the law, for instance. The centre thinks that the law is contained in statutes, statutory instruments and the pronouncements of the highest courts. In a sense that's true. But in many senses it is not. Isn't the *real* law that which *in fact* regulates our lives? And that is an amalgam of our understanding of the law in the books, our appetite for disobedience, and our assessment of what we can get away with.[7]

Both Kingsnorth and Scott suggest ways in which their insights might inform our modern, Western political culture. Real folk heroes, says Kingsnorth, live not in the centre – let alone in Downing Street or the White House – but over the edge:

> Wherever you live and whatever culture you come from, it will offer up at least one folk hero who earned his or her status through state-repelling behaviour. Folk heroes mostly do, which should tell us something about the relationship through history between the folk and the state. Here in Ireland, virtually every historical figure wears state-repellent garb, but in England too we have hundreds of pirates, highwaymen, outlaws and rebels to choose from. You all know the name of the most famous: England's shadow self Robyn Hode, who flits through his shatter zone, the English greenwood. with his merry band of refuseniks in tow. We could do worse than to find our own greenwood and take our stand there, beneath the shelter of its great, ancient oaks.[8]

If there has to be a state – at least to fill in potholes and clear the sewers – edge-people will tend to prefer a monarchy. Monarchs, unlike the Machine, are human. States tend to believe themselves to be invulnerable and immortal. Kings and queens have bad hips and tempers and know they will die. They are like us – or much more like us than is the Machine. A bad monarch is one who has begun to think that they are like a state. That can indeed have nasty consequences, but they are more curable – by the death of the monarch, if not by the wise counsel of a vizier – than the presumptions of a state.

In the Greek Golden Age, to be a good citizen was to participate actively in the body politic. In those days, perhaps government – being government of a small *polis* – was really to do with the thriving of the individuals within it (though, in that communitarian age, individual thriving tended to be seen as the thriving of the body politic of which the individual was an organ). Nowadays, government is to do with the thriving of government and its proxies, and accordingly being a good citizen perhaps demands explicit and radical non-participation in the political process.

It's easy to be intoxicated by all this, and in that delicious anarchic drunkenness (there's nothing quite like it), to forget the rather obvious fact that, for better or worse, I'm not a lean subsistence farmer in the highlands of northern Thailand, but a chubby, middle-aged, middle-class suburbanite, living more or less happily, for some of the time, in a mid-terraced house a stone's throw from the centre of Oxford. I don't want to rip up my social contract, and if I did I'd still want to hang on to many of the benefits it gives. I don't want to rip it up because I like my neighbours, and don't see a clear distinction between their welfare and mine, or indeed between their identity and mine. My shape is determined by the joyous pressure of all the bodies in the community around me. No: I want to renegotiate the contract, to make it clear that my obligations are to those neighbours and not to the Machine; that I pay my taxes because they and I are the beneficiaries, rather than because the money lets the Machine rear more children to eat. I want to shout that the writ of the government stops at my threshold, and doesn't get anywhere near my psyche; that I determine the line of the really important frontiers that cross my world; that when the Machine keeps children working eighteen hours a day in a Chinese sweatshop, it doesn't do so in my name. That isn't what most of us thought we were signing up to in the first place.

Most of the time I can, as Kingsnorth and Scott describe, dodge the state. It doesn't mean dodging the tax man, let alone heaving bricks through bank windows. It involves knowing that I have a soul and the state does not, and that accordingly I'm far bigger than the state, and the state and I can't have a conversation – enriching or disturbing – even if we wanted to.

The soul-lessness of the state brings us back almost to where this chapter started: to the contention that the state isn't anything. It doesn't exist – at least not in anything like the same way as a dog or a person exists. I'm concerned that Kingsnorth's and Scott's anger against the state, and the language in which they denounce it, give the state a status it doesn't and mustn't have. They certainly give it a prominence it loves but doesn't deserve. But there's more. They treat it as if it has a mind and intentionality. It doesn't. 'States are not moral agents,' wrote Noam Chomsky. 'People are ... '[9]

Nor are the state's henchmen (the corporations) moral agents. Corporate personality is a diabolical fiction dreamt up by lawyers, but if we treat the fiction too seriously it will suck our personhood out of us. States and corporations are like some species of ghost or vampire, looking for a real body to inhabit and real human mouths through which to utter their commands. The human host never survives, but the veneer of humanity lent by the host often enables the state or limited company to get away with more than they otherwise would. Tap the host and you'll find he (it's usually a *he*) is hollow. There's nothing inside. If there ever was anything inside, after Eton, Balliol and a couple of Cabinet posts had finished with him, the scraps will have been eaten up by the fiction. Just as you can tell a vampire by the absence of his reflection in the mirror, there is one infallible test for detecting a host: he'll never, ever leave the centre. The directing fiction won't let him. At the edges there's no air that fiction can breathe.

The biographer Edward Marjoribanks once set out to write a life of the great barrister Edward Marshall Hall. Marjoribanks assumed that someone who could tug jurors' heartstrings like Marshall Hall would be a fascinating subject for a biography. Not so. For there was nothing there. Marshall Hall was an empty shell. That was the secret of his success. For the duration of the trial, he hosted the soul of the latest murderer, rapist or fraudster, translating their stumbling defence into mellifluous words. He'd have been a successful prime minister; the state would have gladly used his voice on the *Today* programme. But he couldn't possibly have been a real husband. To whom would his wife have been married? Mrs Faustus isn't to be envied.

The fictionality of states, their necessary centrism, and the fact that they can't stretch to the edges where humans live, have some important consequences for our storytelling. There are lots of big clever books about, for example, the history of England. Don't buy them. There's no such place as England, and never has been. And don't, for goodness' sake, buy any political biographies or autobiographies. They're not about *anything* or *anybody*. If you're tempted, remember the test. Take the book to an edge, open it up, and you'll see blank page after blank page.

After all that, I'm back where I started: an edge-person more or less unsafely on an edge. I can live in Sherwood if I want, even if the postal address is somewhere in central Oxford.

What should we *do*, as creatures living on pieces of ancient land arbitrarily named England? Recognize the arbitrariness. Mock the pretensions of the Machine. Enrage it by breathing without its licence and brewing your own rather than buying beer from one of the state's cronies. Argue against the stupidity of the centre's policy of putting all its eggs into a few baskets, and urge it to adopt the decentralizing strategies of fungi. March with placards against hard-heartedness and short-sightedness and destruction; pay your tax because it's only money, and it'll help real people and because the law says you should and lawlessness is scarier than most laws; be radically kind in everything you do, because that's what it's all about and the state, poor thing, can't possibly be kind; never pay the state the compliment of treating it as if it exists, let alone has any human characteristics; never make the related mistake of appealing to the soul or conscience of one of the state's senior slaves – they have neither; recognize that the state is doomed, and bide your time, weaning yourself off your dependence on it, and learning to be free; know that however sophisticated its sensors, it can't know anything significant about you as long as you are deliberately free and determinedly human, for it has no intuition, but only algorithms, and you are way beyond the comprehension of any algorithm; you, by virtue of your species, are an edge-creature with an edge-address, and as long as you retain that identity, resisting the blandishments of the centre, the state will recoil from you as a vampire from a cross.

If it doesn't recoil, it'll suck you dry.

We've looked at edges through the ages – from before the start of the ages themselves, to post-postmodernity. But what is it like to live on and with those edges? What do we do there? What *should* we do?

We live, we die, we eat, we starve, we are cruel and kind, we worship God and gods. We are dissatisfied with our usual modes of consciousness, and we try to get out of our heads into other places. We seek a connection with place and with others. We believe in beauty and significance, and we write poems and paint pictures as bombs are falling.

Part 2 looks at how we pass our time on the edges.

Passing Our Time on the Edge

6

Touching the Edge of God's Robe

And a woman having an issue of blood twelve years, which had spent all her living upon physicians, neither could be healed of any, came behind him, and touched the border of his garment: and immediately her issue of blood stanched.

LUKE 8: 43–44 (KJV)

AN OLD WOMAN LIVED in a cottage in the middle of a clump of elms. We called her Mrs Bogg. No one had ever heard of Mr Bogg, and when I asked my friend Ted why we called her 'Mrs', he said it was because she was married to the devil, or at least was fucked by him in the graveyard on a Friday. I was seven, and didn't know what 'fucked' meant, but since I was obviously meant to know I didn't ask.

She croaked like the rooks nesting in the elms. She had big proud cats which held their tails high in the air. Ted said that people looked like their animals, but Mrs Bogg didn't look like a cat to me – apart from her whiskers. She looked more like a toad, because she was squat and shuffled, and her face was knobbly with warts. 'She says she can cure warts,' said Ted. 'But she can't. Just look at her.'

The business of the warts was more complicated than that, said Ted's sister, Ellie. If you had a wart, she would take it away from you, but she took it onto her own face. That's why she had such a lot of warts. She could take other things too, and she did, sometimes without people asking. When Ellie's mum got a clot in her leg, Mrs Bogg got to hear about it, and when she did, Ellie's mum was suddenly better, but Mrs Bogg had a bad leg for a month.

She didn't just help humans. When a cow was bellowing, unable to push out her calf, Mrs Bogg was taken with terrible pains that made her bend double and clutch her middle, but the calf just slipped out and the vet was cross because there had obviously never been a problem. And when the stream running down from the moor dried up, and all the tadpoles started floundering and dying in the mud, Mrs Bogg's bowels clogged up and the stream ran again, though there hadn't been a drop of rain.

She always smiled at me, and gave me flapjack with cat hairs on when I went past her cottage to count the rooks. I wanted to ask her about the devil but never found the words.

Not long ago they'd have burned her, and Ellie's mum and the farmer and the wartless ones of the valley would all have come to throw on a log.

Long after Mrs Bogg died, and had gone to wherever it is that people go when they've fucked the devil all their lives, I became a crow. It was on the moor not far from Mrs Bogg's cottage, after a long period of fasting and sleeplessness. I flapped, felt the lift, saw the heather falling away, soared, and looked down on the black teeth of the millstone edge, which seemed like molars from above, though they looked like broken incisors if you were a human walking below. Later I landed, and rummaged with my beak inside a dead sheep, and slowly my wings shrunk and I grew hands again.

Usually, when I tell this story, the listener presumes it was a psychotic episode and, depending on their sensitivity, is sympathetic, or cautiously humouring, or worried. And I, defensively, say something along the lines of: 'You think I'm weird. Not so. If you haven't had an experience like that, it's you who's the strange one. For most of our history, those sorts of experiences were as commonplace as a trip to the supermarket is for us now.'

That's an exaggeration. But there's a grain of truth there.

When humans are depicted in the peerless Upper Palaeolithic cave paintings, they are often therianthropes – animal–human hybrids. Antlers sprout from human heads. They're thought to represent shamans in the process of transforming to or from their spirit animals so that they can gallop or creep through the spirit world on the other side of the cave wall and bring back goods or information for the benefit of the clan. Often the animals in the paintings are missing a hind limb, a forelimb or a whole hindquarter. The missing part is in the other world. For the shaman, the door to that world was a different type of consciousness.

There were, and are, many ways of opening the door: of inducing an alternative way of being conscious. Hallucinogenic plants and fungi may have played a part. So might physiological stress. African paintings show shamans dancing around a campfire for so long, and so energetically, that the capillaries in their noses burst and blood gushes to the ground.[1] You get a profound psychological return for that sort of commitment.

However the state was induced, it made frontiers permeable. Doors opened. The shaman could pass into the body of a bison and across a limestone wall. And then a very paradoxical thing happened. For the experience of porosity didn't produce a conviction of mushy homogeneity. Quite the opposite. Out of oneness came the conviction of particularity. It seems that oneness and individuation are interdependent.

The porosity – reflecting the canons of ecology and evolutionary origins as we know them today – was useful. It perhaps allowed true clairvoyant communication at a distance – handy if you're in the wilds without a phone. In *The Lost World of the Kalahari*, Laurens van der Post told how, when out in the African bush with San hunters, they killed an eland. Van der Post asked one of the hunters how the people waiting at home would react to the news. They already know, came the reply: 'They know by wire.' The hunter tapped his chest, and explained that bushmen had a wire in their chests that brought them news. And so it seemed, for when the hunting party got back to camp, preparations for a great party were well advanced, and the eland song was being sung. Van der Post's credibility is not the best, but there are similar stories from many parts of the world. [2]

At the very least, a permeable frontier between humans and non-humans allowed for real sympathy with the herds of caribou, which meant that you had a good chance of knowing where they were going and where you should stand to shower them with rocks. And it made Mrs Bogg double up as the shoulders of that calf jammed against its mother's pelvis.

Journeys across the edge of this world made us the sort of creatures we are. We need to leave ourselves to see ourselves. The perspective from other worlds gave us ourselves: gave us the right and the impulse to say 'I', which is the foundation of all intention. We still go to other worlds to clarify what we are, and to make sure our intentions are true.

The shamans gave us ourselves as modern humans, and ourselves as our*selves*. They could give these gifts because of their voyages. But what made them voyagers? If we knew that, we might become more effective travellers – and so more functional humans ourselves.

There's a fair amount of agreement across all cultures and all ages about the qualifications of a shaman. Shamans are odd. They are not part of the herd. They may have a wasted hand, or a face covered in nodules. They have undergone an arduous and terrifying apprenticeship, typically involving metaphorical piercing or dismemberment, followed by death and rebirth. How could someone who has not died herself be expected to reassure us when we're quaking with the fear of death? They usually live, like Mrs Bogg, at the physical edge of the community – in a forest hut or a remote and dilapidated cottage rather than a suburban semi.

They are shapeshifters. Upper Palaeolithic shamans became bison to gather useful bison-lore. They were as modern as Darwin in recognizing the fluidity of species boundaries, and simply accelerated the process of evolution, or turned back the clock. Mrs Bogg's medieval ancestors turned themselves into hares (to which we are very closely related, and not so long ago in evolutionary time) to escape the lynch mob. So it's unsurprising that shamans have an intimate, conversational relationship with non-humans: witches whisper to their familiars and the familiars whisper back. Plants tell the medicine man that their petals, made into tea, will save the child with a fever.

There's a big, sterile, technical debate about the relationship between atavistic shamanism and what we know now as religion. I'm not going to enter the debate. I'll just note again the therianthropes in Egyptian, Mesopotamian, Indian, indigenous American and many other religions; point out that the Quran and various Jewish sources agree that Solomon knew the language of the birds[3] – as, apparently, did St Francis; that Elijah was fed by ravens; that a blackbird raised its young in the outstretched, praying hands of St Kevin; that when St Cuthbert of Lindisfarne emerged freezing after a night of prayer in the sea he was warmed and dried by wild otters; that Jacob's ladder joined very different but neighbouring levels of being; that Christ seems to have been an archetypal pierced shaman whose death ripped the veil of the Temple, allowing the passage of all into the Holy of Holies – previously accessible only to the High Priest, and then only at Yom Kippur. Jesus is said to have gone on a journey to and through the underworld, returning to bring great gifts, and that so far as shapeshifting is concerned, not only is he said to be both God and Human, but his post-resurrection appearances are distinctly strange (disciples who'd been at his side for three years walked with him for miles and hours but didn't recognize him until he started to tell them about the institution of the Eucharist). And all that's not to mention the Christian doctrine of immanence, which insists that God – and therefore Jesus – is inherent in everything, from stars to seagulls.

As to the shaman's location beyond social and geographical edges: we see signs in all the great religions. Buddha left his aristocratic family and achieved enlightenment sitting under a tree. Hinduism has always lionized the homeless *sadhu* with his begging bowl. The Prophet Muhammad was rejected by the religious establishment and left it, physically and metaphysically, to found a desert religion.

And then there's the story of Israelite ethnogenesis, which can't – at least for the first three and a half thousand years or so – be sensibly disentangled from Jewish religion. Abraham, a destitute refugee from the far eastern fringe of the Mesopotamian world, had to let Pharaoh sleep with his wife to survive,[4] and had to borrow money to buy a small plot to bury her.[5] His journey made travelling and frontier-crossing the central motif in the story of Judaism. The Hebrew nation was forged in forty years of desert wandering, leaving behind the most powerful nation on earth, backs turned on the centre of things.

Jewish dietary laws are eloquent reminders of the importance to God of edges. Amphibians, which ignore the anciently ordained distinction between land and water, are unclean.[6] They are neither one thing nor the other. Go the way of amphibians and you'd confound variegation. The cosmos would be one big bucket of monochrome. Edges in Judaism are emphasized not to fence humans in but to increase the number of categories in which to rejoice, as well as to say that there are moral places into which itinerants careful of their happiness and thriving should not stray. And that *everywhere* is a moral place.

Even when Judaism appears to be centralized, with the establishment of the Temple in Jerusalem, at the very centre of the very centre – the Holy of Holies – is the Ark of the Covenant, still with the wooden poles for carrying it through the wilderness on the shoulders of footsore porters. God remains portable – ready to travel with His people, pleased not with the blood of bulls or goats but by the bearers' blood and sweat on those poles, and by charity. Try to pen Him inside any words or ideas and he'll burst out of them.

Judaism, like Islam, remains a nomadic, desert religion. Its defining memory is *Pesach*, and at Sukkot Jews live in a shelter through which stars have to be visible, to remind them that settlement and cities are not what Judaism is really all about. The Israeli national anthem, 'Hatikvah', expresses the longing of the Jewish people for a homeland. But where will this homeland be? Jews will find it when they go *ulfa'atey mizrach kadima'* – to the margins of the East.[7] It should always be in a liminal space, not fully of this earth – a place from which there is always a view of the wilderness from which the nation came and where, if it's to be healthy, it must always in some sense be. Each year at *Pesach* the Jewish residents of Jerusalem raise their glasses and join the diaspora in the ancient mission statement: 'Next year in Jerusalem.' *Next* year, not this. It's not an expression of confidence that they will still be in Jerusalem next year, but an assertion that they are not there yet at all. The earthly Jerusalem, with its houses, restaurants, art galleries, and with the army to ensure its physical foundations will not be shaken, is incomplete and its destiny uncertain. It's still a community of wilderness-wanderers. Some Jews say, partly for these reasons, that the establishment of the State of Israel itself is sacrilegious: a usurping of the Messianic prerogative – that Zionism has fatally misunderstood what Judaism is.

In wildernesses you have to keep moving between waterholes. This imperative is reflected in philosophical and theological contingency and discovery. The motion of the Exodus is there in the dialectic of Talmudic study in which, by debate (which respects not only the pronouncements of the Sages but also the contribution of the most callow, beardless *yeshiva* boy), the truth is unfolded. All this hints that 'home', wherever it is, must be understood (like everything else) as a process of rolling edges rather than a fixed place.

Christianity inherited the edginess of Judaism, added lots of its own, and became the edgiest religion of all. The story begins, in historical time, with a child conceived outside the boundaries of the marital rules, and indeed well beyond the reach of all metaphysical principles, in a brutally occupied country. There was no room for the periparturient mother in the inn. The neonate didn't have a normal cot, but was dumped in some kind of animal feeder. His midwives, according to the unbiblical tradition, were livestock.

Soon he and his parents were on the run from his would-be murderers. They fled to Egypt, recapitulating, when they eventually returned, the journey of the Exodus. He grew up as the most provincial of provincial boys, in Galilee – the armpit of the Levant. 'Can anything good come out of Nazareth?' sneered a contemporary cynic,[8] in exactly the tone a Tory MP amongst friends might use of small-town chavs.

He never felt at home anywhere, which was hardly surprising if his real home was Heaven. He was an itinerant, homeless preacher, healer and baptizer. He never got a foot on the property ladder, and would certainly have been shown the door at a typical US evangelical church. 'Foxes have holes, and birds their nests,' he observed, 'but the Son of Man has nowhere to lay his head.'[9] It was true: look at the map of his restless, dusty peregrinations around the Near East. While he was waiting to be arrested by the centrist authorities, he begged his friends to stay awake with him. They didn't. Christianity's supreme moment came when God was nailed to a piece of wood outside the great metropolis. His murder, it's said, had the effect of permitting safe travel for all over the greatest edge – death.

Jesus's life and death were no big deal to the authorities of the day. Yes, he was executed as a troublemaker, but crucifixions were two-a-denarius. Blaise Pascal was right to jot in one of his notebooks that Jesus lived 'in such obscurity ... that historians writing of important matters of state hardly noticed him.'[10] His great popularizer, St Paul, didn't stay in Jerusalem, the centre of Jewry, to trumpet the story. Paul's travels were legendary. And the legend says that the then-centre – Rome – killed him as it had killed Jesus, as the tradition says it killed all but one of the disciples, and as it went on to propel over that final edge countless thousands who said yes to the edge-man.

The early architects of Christian thought left the bright lights of Alexandria and the cities of Syria and went to the desert. The desert again, where so much happens. Take St Anthony, the father not only of Christian asceticism but also of monasticism. Following advice rarely found in the investment strategy of GOP-voting evangelicals – 'Go: sell everything you have and give the money to the poor. Then you will have treasure in heaven' [11] – he moved first to the outskirts of the city and then further out into the desert. He crossed not just the metropolitan boundary but also one of the lines between life and death. He holed himself up in a tomb. There he did battle with demons. Their strategy was revealing: they tried to make him return to the metropolis.

Anthony refused, and stayed in his tomb for fifteen years. When he left, it was to go still further into the wilderness. He found an abandoned fort and locked himself inside for another twenty years, refusing to see anyone, and making very few incursions into the world outside. Attracted by the extremeness of his life, pilgrims came to the fort (people on the edge do attract: it's the north and south poles of a magnet that attract – not the middle). The doors were bolted against them, but they said that they heard the clash of arms and armour as Anthony contended with the principalities and powers that sought to destroy him.

At last, some importunate visitors broke down the door and demanded that he taught them. Reluctantly he did, and monasticism was born. But Anthony knew his place. It is the place of us all. He went yet again to the desert, buckling up his armour and strapping on his sword. There he remained until he died, twenty-eight years later, at about a hundred years old, smiling as he saw cheering saints welcoming him home, but regretting that he had not been given more time to repent.

I'd have listened to what he had to say about the meaning of those terrifying, paradoxical, impossibly demanding, occasionally comforting words in the Sermon on the Mount. Perhaps he knew why the meek will inherit the earth and why mourners are blessed.

There's a clear line of succession between the Mount and the Egyptian and Syrian deserts. The line was broken by the calamitous conversion of Constantine. He made Christianity not just a, but *the* centre-religion. Christianity is dramatically radical or it is nothing. To bring it into the political centre and make it draft legislation and build empires is to kill it. The Sermon on the Mount occupies a wholly different space from the edicts of any emperor. They can't be reconciled. You might as well try to persuade Che Guevara to make his revolution a wholly owned subsidiary of the Carlton Club. St Francis, in kissing lepers' sores and talking to the animals, wasn't left-field at all. He was authentic. That he was regarded as outrageous just goes to show how far medieval Christianity had strayed from the Mount.

Christianity had about three hundred years as itself – as a shrill, profoundly unrespectable, counter-cultural movement – before the Christian brand was bought up and most of its distinctive lines discontinued. But in those years, seeds were sown. Not in throne rooms, but in individuals (some of whom may have temporarily occupied throne rooms or later, parliaments).

There are no Christian centres other than the centre of the individual human. That's one of the messages of the Incarnation – an Incarnation in a particular body, which spent most of its time meandering through obscure backwaters of the Roman Empire. There can be no thoroughly Christian institutions or society, for individual humans, not institutions, are made in the image of God. The Church itself, if it becomes an institution rather than a collection of individuals, loses its Christian character. C. S. Lewis's famous sermon, 'The Weight of Glory', observed: 'Nations, cultures, arts, civilizations' (those often central things and places, which certainly include Byzantium, Rome and the strongholds of Reformed Christianity), 'these are all mortal, and their life is to ours as the life of a gnat.'[12] Movements such as the abolition of slavery weren't really the result of any act of an institution or a society – though the legislative endpoint might have been rubber-stamped by a parliament. They were the result of individual consciences – consciences which had generally been opposed by centre places and centre people for reasons that had nothing whatever to do with Christianity, and everything to do with power and self-aggrandisement.

*

Wise religions embody in their practices their received knowledge or instinct of the sorts of creatures humans are. They know, for instance, that we are wandering boundary-crossers. So they institute pilgrimage. All the great religions do. We've heard already about Judaism's central story – the story of the Exodus – which says that one has to leave and walk in order to claim; in order to be oneself. Jews of all traditions flock through clouds of resentment, and sometimes tear gas, to the Western Wall; Hasidim go to the tomb of Rebbe Nachman of Breslov in Uman, central Ukraine, even in the middle of a war.

The Jews weren't allowed to settle, anyway. The antisemitic world made the Wandering Jew carry his possessions on his back. The most portable thing was the Torah, and Jew-hate forced Jews to be even more People of the Book, and even more edge-people than they might otherwise have been. Shoved to the fringe of Christian society, their contribution to every field of human endeavour has been out of all proportion to their numbers.

Islam has the Hajj, a religious obligation which reminds Muslims that they – like us all – are pilgrims at heart, and that almost all of them live in the diaspora (only about 0.13 per cent of the world's Muslims live in Mecca itself).[13]

India is one massive sacred landscape. An afternoon stroll might take you to the right nipple of Shiva for a view to the Himalaya, and back over his shoulder to your apartment. You can't walk the dog or go to the corner shop in India without going on pilgrimage, which is just how it should be for everyone.

Buddhists go to Bodh Gaya, where Siddartha Gautama, sitting under the Bodhi tree, slashed out of the thicket of delusion which had held him fast. Being free, he began to walk. They go too to sit at the feet of any number of spiritual masters who, though they might be found in a flat on the Central Line, are more likely to be in a mountain monastery just below the snow line – for which you'll need good boots, a long time out of the office and very possibly a yak.

Christians have historically gone to some places that one might (and they might) have thought of as centres – Jerusalem, say, or

Rome – but those places were well outside the usual circuits of their lives, and the journeys uprooted the pilgrims and rubbed their noses in their peripatetic origins. But most of the pilgrimage routes are away from the big centres. Chaucer's pilgrims leave London to go to Canterbury – a peripheral place if ever there was one. One very good reason to take Anglicanism seriously is that its central place and site of pilgrimage is not London but a dowdy market town (population 70,000), dumped on the edge of a marsh and surrounded by sheep and industrial estates. One very good reason to doubt its authority is that the Anglican Communion is actually administered from Lambeth Palace (the P-word should raise our hackles and our suspicions), just across the Thames from the even more sinister Palace of Westminster.

The most popular Christian pilgrimage route today (though it is dwarfed by the Hajj and the great streams of pilgrims around India) is the Camino de Santiago, from many different European starting points to the Cathedral of Santiago de Compostela, hanging on to the far north-western corner of Europe. It is said that the cathedral houses the bones of St James, who was beheaded in Jerusalem and whose relics were mysteriously translated to Spain, to be discovered there in the ninth century. But the vast numbers (nearly half a million in 2023) converging on Santiago from all over Europe are a tiny fraction of those going to other out-of-the-way sites: provincial cathedrals, holy wells, monasteries built to shut out the world that now beats at their doors, and the locations of apparitions. Those apparitions, significantly, tend to be to poor, unlettered people in the sticks – like the schoolgirls who saw the Virgin Mary and the Archangel Michael in San Sebastián de Garabandal in northern Spain in the 1960s – rather than to mitred grandees in the corridors of power.

Then there are the secular pilgrims – or, as I prefer, the pilgrims who don't know they're sacred. The great multitudes who spend untold fortunes on boots, compasses and rucksacks to trudge across the world on foot when it would be far quicker to drive or take the bus. They too go to sites they find sacred, even if the sanctity comes only from the soles of other pilgrims. Their language for describing the sanctity might lack religious adjectives, but it's the same in substance as that of the old women who stumble with bleeding corns up the sacred mountain of Croagh Patrick in County Mayo, reciting the Hail Mary, telling their rosaries and saluting St Patrick, who fasted there for forty days.

The sheer numbers and the passion of these ostensibly secular pilgrims speak volumes.

Protestantism frowned on pilgrimage. It focused on the abuses at pilgrimage destinations: the religious-trinket racket and the sale of fake relics and indulgences. It had theological objections too. It was concerned that pilgrims might think their sweat, blisters and exhaustion could buy them salvation.

I don't know about salvation in the sense the Reformers meant, but the business of walking and leaving behind one's old life and presumptions certainly does good for your waistline, and probably for your soul too. It might not be salvific, but it is restorative. It helps to recalibrate: to produce normative humans. Each step helps to restore the factory settings. It is practical atavism. Of course if one thinks, as those fierce Augustinian Puritans did, that natural humans are irredeemably corrupt, this observation won't strengthen the case for pilgrimage. The Puritans metaphorized pilgrimage. All life was a pilgrimage, they said, and John Bunyan made the idea exciting.[14] It was a dangerous idea, for it implied that the only important things about us are our spirits. Bodies had to be tolerated, because spirits were shipped around in them (what a design fault that seemed to be! Omnipotence was often inscrutable), but bodies gave trouble, and had to be kept on a tight leash.

The traditional understanding of most religions (to generalize wildly but usefully) is that humans are mind–body–spirit unities – holy or unholy trinities. Affect one element and you necessarily affect the others. If there's anything in my thesis about quintessential human edginess, it must take this into account. In healthy humans, the members of the trinity travel together. In a merely metaphorical pilgrimage, the body and probably the mind are left at home to grow flabby and resentful, and to mull divorce from the spirit.

Pilgrimage is a journey back; back to what we were and really are. It remakes hunter–gatherers from CEOs and actuaries. This means it is and must be a journey away: away from the physical metropolises where most of us live, and away, most significantly, from that most polluted of metaphysical metropolises: the Self.

If you know that it is crucial to go back – so crucial that you have to take time off work or ignore the pain in your hip or buy a plane ticket to Mecca instead of a new kitchen – you've already thumbed your nose at one of the centre's most revered, demanding and pernicious cults: progress. The cult doesn't survive outside the centre, and backwards isn't an option for progress. If you've bought your boots to go back, you're halfway to escaping the cult.

All the great religions seem to be characterized by great innovation: by radical departure from the previous devotional centres. (Hinduism may or may not be an exception – its roots are tremendously ancient and hard to trace.) Yet the human agents of this innovation usually didn't want to rupture traumatically from the old order. The Buddha didn't intend to found a new religion; he just wanted to see clearly and be free. The foundation of Judaism, if the Hebrew Bible is to be believed, was very much on Yahweh's own initiative: he did the choosing of the Chosen People. Neither Jesus nor Paul wanted to start something separate from Judaism, and neither thought he had. Islam arose when the Jews and Christians of Arabia wouldn't assimilate the Prophet Muhammad's revelations into their worldview. Luther hoped to stay within the Catholic church.

Whatever the hope or the intention, seismic religious change involved relocation: the court to the forest; Egypt to Canaan; Jerusalem to the wider Greek and Latin-speaking world; Mecca to Medina; Rome to Wittenberg and Geneva; and then, since the Bible meant whatever each preacher said it meant, to a multitude of pulpits, first in northern Europe and then to megachurches in Seoul and Tennessee.

Unsurprisingly, since newness *is* the crossing of edges, and the negotiation of edges is the core business of religions, history suggests that if you want to start a new religion, it is shrewd to move across geographical frontiers; to embody in your new geography the philosophical and theological change you want to effect. Move bodies and it will be easier to move minds, souls and spirits. A physical shift will also mean it's harder to backslide; harder to be re-contaminated by the counter-reformers.

There have always been wannabe messiahs and aspirant new religions. The stories of most of them are tawdry – to do not with building houses on rock but with getting one's rocks off. But if they've enjoyed any success at all, they've followed the blueprint: innovate; isolate; move.[15] They're authentically edgy, whether their religion is benign or toxic. Think of Jim Jones, who moved his Peoples Temple from Indiana to the Guyanese jungle. In November 1978, 913 of his congregation, including over 270 children, either committed suicide by drinking punch laced with cyanide, or were shot or had their throats cut by their co-religionists.

Few deliberate religious innovators left anything meaningful behind. One who did was Akhenaten – born Amenhotep IV – the heretic pharaoh of Egypt's Eighteenth Dynasty. Some say he invented – or discovered – monotheism. What he in fact did was outlaw the worship of the old gods and announce that there was only one god: the visible sun-disc, the Aten. Before Akhenaten, Egyptian religion had increasingly focused on the sun and on the sun-god, Ra. Ra's synthesis with the local god of Thebes, Amun, was known as Amun-Ra, and by the time Amenhotep IV came to the throne, Amun-Ra was the king of the gods in all or most of Egypt.

Amenhotep IV's revolution was not to assert the importance of the sun – that was already amply recognized – but to declare that there was one god, not many; that this god could be seen and felt without the mediation of priests, without mumbling formulae or fumbling with amulets; and that he, Amenhotep, was – as his new name, Akhenaten, announced – the way the Aten manifests in the world; the effective agent of the Aten. It was a vaunting personal claim. Since the sun shone on all, there was a hint of egalitarianism, but his creed was a long way from the priesthood of all believers.

Akhenaten started, like Luther, by trying to change the centre, and at first did not outlaw but merely demoted the old gods. At the start of his reign he began a huge building project at Karnak, the epicentre of the Amun cult. The old temples were dark and smelt of blood. Aten's were open to the light and smelt of flowers. There were not even lintels over the doors, for nothing should obstruct the rays from which life and enlightenment came.

This co-existence of Amun and the Aten did not last. It is not clear why. Perhaps the priests of Amun cut up rough. Perhaps Akhenaten's own piety deepened. But in the fifth year of his kingship he changed his own name, banned the traditional gods and their lucrative and entertaining processions and festivals, scratched their names from the walls, and made plans to relocate his court to a brand new city in pristine desert, about equidistant between the southern metropolis of Thebes and the northern metropolis of Memphis. It was a physical and theological statement of shattering significance. 'No man appears to have made a greater stride to a new standpoint than he did,' wrote the archaeologist Flinders Petrie.[16]

Akhenaten called his city Akhetaten – 'Horizon of the Aten'. It is known today as Amarna. He may have chosen the site because at dawn the sun rises precisely in a gap in the long rampart of limestone cliffs which climb steeply to a desert plateau and form the eastern boundary of the city. The sun resurrected the dead who had been sleeping at night in their tombs. Because of the breach in the rock, Akhetaten's dead were woken earlier than anyone else for miles around.

When he left Thebes for Akhetaten, Akhenaten resolved not only that he would never return to Thebes, but that he would leave Akhetaten. It was logical enough. If this was the uniquely consecrated city, and he was the uniquely consecrated king, where else could he manifest fully the power of the sun-disc? So far as we know, he honoured the promise during his lifetime. He turned his back on the world's most glittering culture – the nearest thing, indeed, to the very centre of the world. Under his father, the august Amenhotep III, Egypt had become wealthier, statelier and more revered than it had ever been. Its temples were vast, and encrusted with gold, silver, bronze and jewels. Its neighbours knew when they were beaten, stopped rattling their sabres, and rattled instead the chests of rubies brought to Thebes by their sycophantic ambassadors.

Akhenaten left Thebes for Akhetaten on a royal barge. A modern felucca – if it didn't use an engine – would take longer. It wouldn't have slaves to row, and would have to rely instead on the hot wind from Nubia, brewed in the highlands of Ethiopia. When I've gone from Luxor to Al-Minya, forty miles downstream of Amarna, I've usually gone by train. It takes seven shuddering hours to cover the two hundred and fifty-odd miles. It seems longer. Akhenaten's journey to Amarna took about three days. In the resentful, outraged eyes of the priests of Amun-Ra, it was a journey back to the unblessed time before the origins of dynastic Egypt. The king had long been the high priest of the sun, and so the guarantor of cosmic and temporal order. With the king's departure from the true temples, order was threatened. They were proved right. At least for a while.

Nothing much has changed since Akhenaten came this way.

From Luxor the train wheezes north along the course of the Nile. When it stops at a station or a crossing, or just because the driver wants a pee or a bag of dates, it gasps until it goes again, and when it goes it snorts, and bee-eaters pipe round it, picking off the flies stunned by the snorts.

To the east there's that rock ridge. Beyond it, the desert howls. To the west, after a thin strip of green – a strip that, economically, *is* Egypt – there's nothing (if the desert is nothing) until the sand spills into the Atlantic.

All life crowds into the green: jostling, calculating, planning, arguing at high volume; trying to etch out a life, make the shape of that life lovely, and never despairing. Here in the strip children learn stories, and learn that the stories are true in a different kind of way from the truth of the *fuul* they get for breakfast, dinner and, if they're lucky, lunch. Here is the cheating, the faithfulness, the honour and the big-heartedness of Egypt. Here men sit on the pavements, maggots wriggling in their sores, and here the bilharzia parasites are already burrowing through the skin of those kids swimming around the dead donkey, and squirming into their bladders. Here the avalanches of used nappies cascade into the canals, and the egrets look ashamed.

The prayers trickling from the mosques are the prayers of a desert people made sedentary. The Prophet's first successful appeal was to the Bedouin, who knew the imperatives of sun and grit; whose armies swept through Arabia, the Levant, North Africa, and on into Europe, carrying with their green emphatic flags the irresistible reminder that all humans come from the edges, like the Prophet's soldiers, and would return there. A reminder comes from the minaret five times a day, but the people barely need it. They can see the edge: it begins just beyond their patch of *molokhia*. They carry their friends and fathers (fast, because they quickly start to smell in this heat) to the edge beyond the edge.

Men splash merrily in an outsourced bit of the Nile. In the frame made by the train window, all that's changed in the last 2,500 years is the route of the canal and the fact that the splashers' pants are made by slaves in China from oil, rather than by their wives at home from cotton they grew themselves.

I have to stay overnight at Al-Minya before doubling back to Amarna. In the dark, by the Nile, there's an ululating wedding, fluttering with eyelashes as false as pharaohs' eyebrows. I'm not sure I'd want my wedding party on the *Titanic*, however merry its fairy lights and however cardiac the thump of its disco.

In the morning, George, who's going to drive me to Amarna, spends an energetic half-hour arguing with the police. They want to escort us all the way there and all the way back, which would slow things down mightily. It's not so long since this part of Middle Egypt was a hotbed of terrorism, and they're still jumpy. George, a great diplomat and advocate, manages to shake them off, and we're away on the desert road. Flaking concrete and fly-blown breeze-block huts give way to sand. We drive past a desert necropolis – Muslim and Christian cemeteries next to one another – the Muslim one far larger than the Christian. They have proper houses for the dead; they make up a big, comfy, affluent suburb of Al-Minya. There are cement mixers and trucks busily extending the suburb. There's nothing for them to do in Al-Minya itself, but death's always good business.

This isn't sweeping, rolling, cinematic desert, but a vast sandy rockery, with rusting bulldozers rather than gnomes. The river is the big presence: a fringe of distant palms, or a line of haze more succulent than the heat haze above the rocks, or a slightly cooler right cheek as we drive south.

Amarna has a good claim to be the most pivotal place in history, and it's pleasing that you'd miss the turn if you blinked. The sign has been scoured with hot sand and used as target practice. George, of course, doesn't need the sign, and soon we're rumbling up the unmade road to buy a ticket and collect our guards. The guards will go with me everywhere, shambling five paces behind in their too-big trousers and too-tight camouflage jackets, fingering their triggers and watching not for signs of homicidal fundamentalists but for signs that I might be taking out my wallet.

The ticket seller is astonished to see us, and makes us sit down, drink mint tea and try to explain why on earth we're here. Last week there was a Swede with a beard and a bowie knife, he said, but it's really not the time of year. As we leave the office, he looks at us with wonderment, shaking his head.

The tombs are what everyone (sorry, no one) comes to see at Amarna. And for anyone fresh from the Valley of the Kings they are at first a disappointment, then a puzzle, and then – after a week's or a month's or a lifetime's reflection – an epiphany like nothing else from the ancient world.

Like the city itself (which was erected fast, using blocks far smaller than those used for previous royal buildings), the tombs, burrowing into the rock escarpment above the plain, are jerry-built. The king was in a hurry to make his point: to create his sanctuary; to erect the boundary wall separating him from the old ways. Plaster was slapped onto poor-quality stone, and the paintings on the plaster are often crumbling into the wind. But what paintings they are! This is something new. It's as if the rays of the Aten have thawed the frozen figures of earlier Egyptian art. The humans unbend and shrug off the convention and sclerosis of the millennia; horses flow; birds bask in the sun, ready to fly. Everything is more fully itself than it had ever been.

I've had two comparable experiences. One was walking through the sculpture gallery at the National Archaeological Museum in Athens. There are lots of very similar *kouroi* statues there, arranged chronologically from oldest to youngest. If you walk briskly past them, with half-closed eyes, it's like flicking through one of those books that make the picture move. The *kouroi* unfreeze. Their shoulders unknot, their arms move slightly from their sides, and their smiles climb from their lips to their eyes.

The other experience was watching someone come round from a general anaesthetic. She was as white as the bedsheets, but colour rose first in her face, then in her wrists and then bloomed in her fingers, which started to tremble and clench. Her eyelids flickered as if the eyes beneath were watching a car chase, and her lips worked, commenting on the chase, and then her eyes opened, blindly at first, then fixing hungrily on all the solid things in the room. Orthodox Jews, first thing in the morning, thank Hashem for restoring to them the soul that had been absent and drifting through the night. That's how it looked to me at that bedside. In those tombs at Amarna, it seemed that the soul had come, to Egypt or come back.

The Aten radiates from these murals, making everything possible. At the end of each ray there's a hand. In the cruder representations you see only a finger and a thumb. The sun is *intentional*;

it's manipulating the world. In finer murals, whole hands are visible, and they seem to stroke.

Akhenaten always looks odd. He has an elongated, horse-like face; fish lips; pendulous breasts; a pot belly; big, breeding hips; and long, thin, skeletal fingers. The literature is awash with speculation about this appearance. Did he have some medical condition – perhaps Marfan syndrome? Were the strange features allegorical? Were they telling some story about his style of kingship or his relationship to the Aten? Did the hips suggest that he gave birth to his people, or to a new idea? We don't know, but the other figures in the Amarna paintings – including his wife, the notoriously lovely Nefertiti – share the same unusual characteristics, though not to the same degree. Perhaps the message was that everything took its form and its being from him, as the mediator of the source and sustainer of life?

Since the rays of the Aten touched everything, everything – from king to fish – tasted divinity. Swam in it. And the most lasting and explosive insight of Amarna art is the sense of the overwhelming significance and sanctity of the everyday. There are real human relationships here – not the artificial relationships created by status and title – and they are valued more than the posturings of monarchy that characterize traditional Egyptian art.

Akhenaten is usually seen with Nefertiti. He plainly loved her very much and valued her very highly. In the paintings she's more or less Akhenaten's size – not one of the little doll queens of the Theban necropolis, shrunk in the murals to show their insignificance compared to their great lord. The children are smaller, but not grotesquely so.

Stepping into the Amarna gallery at the Egyptian Museum in Cairo is like stepping from the press and hoot and smog of a seething Cairo roundabout into a sedate, air-conditioned apartment on Zamalek Island. The faces are quiet, calm, and the smiles gently satyrical. Those full lips on the great statues of Akhenaten are capable of the thoroughgoing scepticism possible only when you're really certain of a few things – things like the daily rising of the sun and its significance.

The most moving piece in the gallery is a small, coloured frieze showing Akhenaten and Nefertiti not smiting Egypt's enemies or using their slaves as footstools, but playing with their daughters. The parents play with the children not just to make a political point. It's not one of those press photos of the happy family in the Downing Street garden, designed to win votes. They are wholly gratuitous pictures of a happy family being happy. There's nothing distinctively royal about the family groups, unless family is by nature royal.

This is a radically new kind of kingship, which knows the supreme value of the things made possible by the sun: bread, play, family. Of course there are court politics and aspirations, but this looks like politics in the service of life, not life in the service of politics. Amarna is a festival of the humdrum whose strapline is that nothing is humdrum. You need to be on an edge to see that. Most of us – at least in our better moments – know that we're always on the edge, and today we don't need to be reminded that the centre has lost the plot. Look at those polls showing disillusionment with politics and politicians – whether left-wing or right-wing or anything in between.

Everything about the city is a repudiation of the old centrist ways. It's barely a city at all, in fact, although 'it' is huge. It takes George fifteen minutes, on empty dirt roads, to drive from one end to the other. It rejects the grid layout that encoded the old, bureuacratic, immovably hierarchical society. Instead it's a chaotic constellation of tiny communities scattered across the plain. Devolution and self-sufficiency were clearly important principles, for each community had its own workshops, grain silos and animal pens. The self-sufficiency was almost manic. Very unusually for this period, many communities had their own wells. It's as if Akhenaten's need to assert the all-sufficiency of the sun required him to declare his independence from the Nile itself.

For me, the best sign of the radical newness of this reign is on the walls of the Amarna tombs. The artists took conspicuous care with Nefertiti's *ears*. I think she was a gifted listener. The Egyptian royals weren't noted for that.

Akhenaten's own tomb is very like the traditional tombs in the Theban necropolis, except that it's a family tomb. He seems to be expounding the old order to itself: saying that it can only be understood properly in the context of the Amarna philosophy – as Christian supercessionists claim that much of the Hebrew Bible is really about Jesus.

Akhenaten was no St Francis. He thought he was special, as his name tells. There is no real or false modesty in Amarna. But to my eye, Akhenaten's self-belief connotes duty rather than despotism. His sense of his own significance doesn't feel self-aggrandizing. The tomb murals suggest *noblesse oblige*.

The centre hates and fears the edges. We'll look at that in Chapter 15. Hate and fear are seen dramatically at the end – or in a pause – of the Amarna story. Akhenaten died in the seventeenth year of his reign. It's not clear how. But whether he died of natural causes or was a victim of the old guard, the old guard tried to forget he had ever existed. They didn't let him rest in his tomb, but dragged his body back to Thebes, trashed his city and obliterated his face and name. The ruined temple of the Aten at Amarna is now a temple of the squatting dog and the hitched *galabiya*. You have to pick your way carefully between the turds.

Akhenaten was succeeded by the famous boy king, Tutankha-*mun*, who may well have been Akhenaten's son and who, if so, began life as Tutankhaten.

Who really won?

Akhenaten, insisting on his role as sole agent of the Aten, had been more autocratic than many previous pharaohs. The reaction against this view of kingship diluted the power of the monarchy and caused a shift towards theocracy and a more direct, peasant-empowering relationship between Egyptians and their gods. I suspect Akhenaten would have thought that a triumph – a desirable stage in the evolution of religious consciousness. The sun, after all, shone on all.

Perhaps he won the theological argument too. Religious thought consolidated around Amun, who was seen as the key god – if not *the* god. The other members of the pantheon came to be seen as aspects of Amun. It had been moving that way for a while, but Akhenaten accelerated the process.

As the Aten dropped into the sand beyond the Nile, I walked back to the car, where George was asleep. The call came loud from the mosques of the nearby village of El-Till: 'There is no God but God.' It took a while, but Akhenaten triumphed in the end.

When I got back from Amarna I went to stay, as I always do these days, in a guesthouse in Islamic Cairo. Just round the corner there's a Mamluk mosque, and the caretaker was happy for me to sit there for a couple of hours a day.

This mosque straddles many edges. New is bolted to old, and after a season of Saharan dust and Mediterranean rain, the join is invisible. Mostly there's only the murmur of an old man reciting the Quran to himself. But at four in the morning there's an ear-rupturing reminder that it is better to pray than to sleep, and at four other times through the day, at equal volume but in different words, we're reminded where our priorities should be: that we're not just bio-logical; that at least at these times, flesh should give way to spirit.

The old light of the sun, which has taken eight minutes to get here, mixes with the old dark in the corners of the graves of saints and tyrants. Children lean against the graves and eat their sandwiches. Birds fly in and out and lay their eggs on nests made of bone-dust and rice-pudding boxes. This coming together of time and space and life and death and sensation and silence produces the amalgam we call peace, poise and dignity.

This has been a chapter about religion. It's unfashionable to write about religion, except as a political force. Much of it has been about *the religions*: it's even more unfashionable to write about that. But getting out of your head? That's always popular.

7

States of Mind and Un-mind

I have found both freedom and safety in my madness; the freedom of loneliness and the safety from being understood, for those who understand us enslave something in us.

KHALIL GIBRAN,
The Madman[1]

'MY JOB,' DRAWLED THE psychiatrist, sloshing Rioja round his gums, 'is to put people back into their right minds.'

'Their right minds?' queried the philosopher, who was the evening's host. 'You're making a lot of assumptions in those three words. I wish I could be as confident about anything in my professional life.'

'Why would you want to do such a thing, anyway?' asked the writer, who had a colourful psychedelic past. 'Right minds are terribly boring. They've never done anything interesting or significant.'

'That's easy for you to say,' the psychiatrist replied. 'You've not seen the pain of depression or the terror of florid psychosis.'

'Nor, I dare say,' interjected the theologian, 'someone tormented by what he thinks – and for all I know he may be right – are demons.'

'Drink up, anyway,' urged the philosopher. 'I much prefer you all when you're not in what your employers regard as your right minds. At least you'll start telling the truth. And if it's only *in vino* that we have *veritas*, perhaps we should pump alcohol into the public water supply.'

They all had a point, but the truth of the matter didn't become much clearer as the evening went on.

The wine-god, Bacchus, who increasingly possessed that dinner table, has an ambivalent relationship with the truth. He doesn't seem to be interested in the truth for its own sake, and whenever truth and goodness form an alliance, he tends to throw in his lot with badness. He is, however, very interested indeed in scrubbing off pretensions and incinerating certainties. In antiquity he was associated with sexual ambiguity and gender fluidity. He's more of a destroyer than a creator, but might sometimes clear the way for something creative. When he turned those prim Athenian housewives into screeching, dismembering Maenads, he showed them not what they were but what they were not and what they might yet be. He scraped the scales from their eyes and they saw new things. No doubt their kitchens were much more interesting when they returned gore-stained from the mountain.

A good deal of the history of humans (and the most influential part of that history) has been concerned with dissatisfaction with our usual states of mind and ways of being. An evangelical song insists: 'There must be more than this.'[2] It's a good executive summary of our general outlook from the last Ice Age onwards. This conviction spurred the shamans across the cave wall and into the bodies of wolves; ignited and maintained religious and political instincts of all shades; propelled thrusting executives up greasy poles instead of spending time with their children; made the grass on the other side perpetually greener; created *The Divine Comedy* and the territorial ambitions of emperors; bore Machiavelli, Michelangelo and Mother Teresa; and is responsible in probably equal parts for much of human ecstasy and misery. It makes us smoke tobacco and many other substances, inhale gases that help us to taste reality rather than merely see it, drink tea and coffee so that we can be more awake (for what, exactly?), sit in the lotus position in cold rooms and watch our breath until the air starts to dissolve our self, drink beer to forget the week just gone, take plants and mushrooms which slacken off the reducing valve of our brain and let more data flow in, supplementing the usual daily dribble that has been set as our default allowance for, perhaps, a small part of our history. We ride motorbikes too fast, climb hills, pound the streets in lycra to change the shapes of our bodies and get a pulse of serotonin, cuddle our partners because it increases the secretion of comforting oxytocin, pursue orgasms and foxes and promotions and education and insight and big waves and the past and the future and God and the devil and an allotment and a holiday in Torremolinos and a fast, shiny car and a Tory-free utopia and a land flowing with milk and honey even if it is in Basildon and a child who can play the oboe and a place in the Premier League for your tribe and peace on Earth and goodwill to all men, and indeed anything at all other than what you've fucking got at the moment.

Where would I be without my morning coffee or my evening wine? The answer, I suppose, is 'myself', and that is the least desirable location in the cosmos. I'm desperate to leave it. Desert people dream of hanging gardens and flowing streams; English public schoolboys from the Home Counties dream of deserts.

The arts are obsessed with newness – even if the new is palpably worse than the old, or an iteration of the old. If you want to lure the Berlin Philharmonic to your town, they'll have to play something unusual. It doesn't much matter that it's musically maladroit and painful to hear; newness is the thing. Ask them to play the Four Seasons or Beethoven's Fifth and the answer will be no.

Whenever we see an edge we want to cross it. If there aren't any obvious edges to cross, we imagine there are, or we manufacture them so that we can get on with the great human mission of edge-crossing. It's only natural. And, since we are naturally supernatural, it's only supernatural too.

The powers that be, and the academic establishment, want to keep us inside our own heads. The powers like it that way, because we'll be easier to control. They'll know where we are and what we're thinking. We'll be mappable and calculable. Hence the war on edge-crossing and edge-crossers (we'll smell some of the napalm in Chapter 15). The academy wants to study and promote ordinary consciousness because that's the sort *it* handles best. Plato, of course, saw philosophers as the top of the ontological and societal tree. Though he himself had a mystical bent that would bar him from tenure in any decent modern university, what the Athenian academy meant by philosophy was, by and large, the sort of cogitation you do with ordinary, everyday consciousness. It was, well, *cognitive*: systematic, syllogistic; about the orderly procession of propositions; about going from A to C via B, being able to demonstrate every step of the route, and travelling always in a dead straight line.

The word 'dead' in the phrase 'dead straight line' is revealing. For that's not how real, live humans do their travelling. To insist that the type of consciousness that proceeds this way is the best, let alone the only, way to do thinking is pathological. It means ignoring a huge body of evidence about how we do science and ethics. It means ignoring most of what we know we are. We are icebergs: most of what we are (as every psychoanalyst, every readable novelist, and every child knows) lies well below the surface of our workaday consciousness. Why do I do what I do? I can concoct a reason palatable to my own cognition and to any philosopher who might be listening, but it's unlikely to be accurate, and it certainly won't be completely accurate. It's likely to be a more or less facile rationalization of the real 'reasons' – which are dictated by (depending on your choice of language) Jungian archetypes, Freudian impulses, God, gods, hormones, the weather, fear, hope, an undigested lunch, or the motion of the stars.

We live and think, in other words, across the edge of our ordinary consciousness. Perhaps that's why we're always so desperate to leave behind our ordinary consciousnesses and the places associated with them. We want to go to where the action really is. This would accord with what we've already seen of our history and our nature: we are natural psychonauts, adept at navigating between different planes of being.

Voluntarily and involuntarily we are always leaving the only realm of consciousness upon which most philosophy is based. That's not usually by being caught up to the heavens in a mountaintop epiphany, or entering the lush, green, jaguar-prowled world of ayahuasca, or even the merry, hazy, bucolic meadows to which a couple of pints of farm cider will take you. It's more usually by falling asleep.

Our dream worlds are more interesting than the office. They draw on influences much older, more formative, and more charismatic than those listed in our CV. If we sleep for eight hours a day sleep takes up a third of our life – but much more than a third of our living happens there. We've bought into the philosophers' lie, and weigh our worth and our achievements simply in terms of what we do in our waking time with our waking consciousness. We discount our sleep. If we live for eighty years, we'll have been asleep for about twenty-seven of them. Probably most of our life's real work will be done in those twenty-seven. It is not for nothing that psychoanalysts are obsessed with dreams, or that the god Asclepius healed only when the supplicant slept and dreamt in his temple, or that, in the august tradition of prophetic dreams, we seem able to slip over the boundaries of everyday consciousnesses, space and time.

We can learn to navigate more confidently in the lands of dream; to inhabit those twenty-seven years more mindfully; to direct our dreams. There is a well-established tradition of lucid, wakeful dreaming. It has been cultivated particularly in Tibetan mysticism. Though it can be learned, about 20 per cent of us have at least one spontaneous lucid dream every month. These are potently multi-sensory dreams – like using a really sophisticated virtual reality system, but about a thousand times more intense. Though the dreams are convincing, we know we are dreaming, and can take control.

Try to enter one when you are next in the borderland, called hypnagogia, between what we arbitrarily call waking and what we arbitrarily call sleeping. Let the hypnagogia have its way for a while. Float in it, paying attention to the whole environment, but not fixing on any particular point. The borderland will slowly draw you into itself, and then you'll be able to use your own feet, and walk around, and be an agent there.

We can engage more consciously with our more usual dream

worlds too. Dreams tend to bolt when we open our eyes, making it hard to trace their effect on our day. It need not be so. Simply keeping a dream diary seems to integrate dream worlds into the rest of our lives. There's a general rule in play here: if we pay attention to the edges that course through us, they honour that attention. Real self-knowledge can begin with a notebook and pen on the bedside table.

I am slowly learning to walk – just with stumbling steps at the moment – in the borderland.

One winter I was ushered brutally into those lands, sitting in a freezing wood in Derbyshire, listening to the screams of dying things and trying to stay awake because I thought that if I did, something would happen.[3] I failed, and my failure – as so very often with failure – was much more creative than the success I had planned. I'd wanted something to happen. It did. My head fell onto my chest, and as it jerked back up, it was as if a switch had been flicked, throwing open a trapdoor between my consciousness and what lay immediately beneath. I fell through, into a scintillating wood – rather like the one I'd been in, but *more so*. The leaves trembled with light, and the light tasted of bubblegum and the trees smelt of hugeness though they looked no bigger than normal trees. My thumbs were colossal but appropriate. I could have crushed any of the trees and held the scent of hugeness to my nose, but that was unthinkable because I had no idea where I ended and the wood began, and because it would have been uncharitable, and charity (which rose from the ground like hot mist) made my hands warm, though my body was still shivering from the less-real wood where 'I' – whatever that was – had been sitting. I could see 'my' body, and pitied it, and was sorry that it was cold, and grateful to it for all it had done, but it was a disinterested pity and a distant gratitude. There were figures, some of them sexually interesting, but I had no desire to possess. They all looked young, yet they were already old when the first plankton clouded a primordial sea.

I knew it wouldn't last, but I knew too what 'I' had to do when it ended. I had to beg Sarah for forgiveness; I had to rip up that argument and start again; I had to re-cast that sentence; I had to finish that bedtime story for the kids; I had to go again to a beach in the Faroes where I had been too distracted to be happy, and to a mudflat in Mauritania where I'd got the wrong end of the stick, and to a conference on a Wadden Sea island where I'd been presumptuous, and to a village in the highlands of West Papua where I stupidly thought I'd understood, and back to the Danakil Depression to know that there was no need to be afraid, and to a cider house in Bristol with Adam to put things right, and up the summit path to the right, and not to the left, on a mountain that looked over to the Isle of Rúm.

All this specificity alarmed me. It seemed like manipulation, though it didn't feel like that. It introduced me to the distinction between seeming and feeling. I'd have been much happier with a fuzzy sense of well-being. It was, anyway, quite something for the couple of seconds it must have lasted. I opened my eyes, and the 'real' wood seeped back; thin, pale, etiolated, dying, dilute. If the pub had been open I'd have gone there and drunk too much beer for reasons I couldn't have identified.

I still have to have that talk with Sarah and with Adam, and the argument looked OK to me in the morning. I re-wrote the sentence (though the first one was better, said my editor), and finished the kids' story (they resented it and wanted to watch *Outnumbered* instead), but on the Faroese beach I found a broken-winged gannet in great pain, and killed it, and on the Scottish mountain a man was crying, and I gave him some chocolate. I'll be back in Mauritania and the Wadden Sea too, but the work I should have done in the Danakil and Papua will have to be done in England.

Others have brought back much greater riches from the hypnagogic edge-lands. Aristotle knew of these lands, though it is not clear what debt he owed to them. Without them we may not have had Beethoven's *Eroica*, any of Tesla's gadgets, the Ring Cycle, or the Waverley Novels. Edgar Allan Poe would not have been a high priest of the gothic unless he had visited regularly, and it is not fanciful to think that Charles Dickens' extraordinary imaginative fecundity was rooted there. *Oliver Twist* contains a description of hypnagogia that will ring bells with those of us who have nodded off and nodded on in an overheated lecture theatre. Oliver describes a kind of sleep which 'holds the body prisoner', 'does not free the mind from a sense of things about it', and enables it 'to ramble at its pleasure'.[4]

Perhaps the main effect of hypnagogia is to expose and expound the connections between parts of our thoughts and parts of ourselves that our usual type of consciousness insists are unrelated. That usual consciousness is highly conservative. And it's a fanatical filer. It creates and maintains divisions between domains that are meant to inform one another. It loves edges – but only the edges it has asserted – and regards as criminal any commerce across them, coming down on any smuggling with the ardour of a border officer on commission.

The officer has no jurisdiction in the hypnagogic zone. Ideas flow promiscuously, cross-fertilizing one another. Great things can be spawned.

Salvador Dalí often went to sleep holding a big key over a metal bowl. As he dropped off, the key dropped too. The clang in the bowl woke him, catapulting him into the edge-lands. There he found the images we all seem to recognize: images that are strange but strangely familiar.

Edison had a similar practice (he slept with an iron bucket on his lap and a coin or a steel ball in each hand), and attributed to it much of his scientific iconoclasm. A version of the story about Newton's gravitational ponderings claims that the famous apple fell on his head when he was asleep. It is almost certainly untrue, but if it had been, Newton (known to be prone to hypnagogia) might have been ushered into a hypnagogic epiphany that would have saved him many hours of tedious equationizing: the solution might simply have been handed to him in the margins of the mind by some generous emissary.

Leonardo da Vinci claimed not to sleep more than two hours a day. He had lots of naps. Lots of naps mean lots of edges between sleep and non-sleep. I wonder if he saw there his flying machines, his siege-engines, the gusts of blood around the heart valves, and the smile of the *Mona Lisa*.

The early stage of sleep, in which hypnagogic experiences can occur, typically lasts only a few minutes, and amounts only to about 5 per cent of a typical night's sleep. It's a true edge: a narrow door. We then drift into a deeper substratum of ourselves. But before we dive deeper, much can be achieved.

A 2021 study invited its participants to solve a mathematical problem – new to each of them. Subjects who had spent 15 seconds in 'N1' sleep – early non-rapid-eye-movement sleep – had an 83 per cent chance of discerning the rule that would crack the problem, compared to 30 per cent of the participants who stayed awake and 14 per cent of those who went into deeper sleep.[5]

What might we be and what might we understand if we could spend more time at the margins of our selves?

Halfway between Cairo and Alexandria, in Wadi Natrun (where ancient Egyptians got the salt for pickling mummies), is the Coptic desert monastery of the fourth-century St Pishoy. I was shown round by a black-robed monk who had a PhD from Princeton and spoke seven languages, including Japanese.

'Here,' he said, 'is the powerhouse of the monastery.' He took us into a tiny room. There was barely room for three of us to squeeze in.

'St Pishoy used to pray here,' said the monk. 'He fastened a rope to this chain, and tied the rope to his hair and his hands so that if he fell asleep he'd be jolted awake. You might think this monastery is built of stone. It's not. It's built of those exhausted prayers.'

No doubt this sort of devotion to duty is rewarded. In St Pishoy's case sleeplessness, which is supposed to be bad for us, made his body incorruptible. It lies immaculate and fragrant sixteen hundred years after his death. I wonder, though, if at least part of the reward is in the coin of hypnagogic spiritual experience.

Most religious traditions have an ambiguous relationship with sleep. It is a realm in which, in dreams, visions can be delivered unfogged by the obfuscations of ordinary consciousness, and truths enunciated without the distracting noise of our churning minds. Yet sleep is often despised by ascetics. It is lumped together with goose-feather pillows and silk sheets as a triumph of the sensual body over the spirit. And of course the bed is the throne of slavering, soul-eating Eros.

Being able to live for long periods without sleep indicates, in many traditions, that the body has been conquered and that the believer lives on a different plane – like the saints said to survive on nothing but air or, in the Christian tradition, Eucharistic bread and wine. Fasting from food and fasting from sleep are ubiquitous hall-marks of spiritual elitism. St Pishoy is in good – and by no means all Christian – company.

The famous stylites of the Syrian desert, who lived on the top of pillars, can't have had much unbroken sleep. When St Jerome felt the predatory approach of sleep, he smashed his head on the ground. St Catherine of Siena had a short nap on alternate nights. St Macarius went without sleep, it's said, for twenty days at a time. Many other great spiritual warriors similarly went sleepless, or slept on the ground, or had rocks for pillows.[6]

In the more austere Zen meditation halls, where the sitting medi-tation sessions can last for many hours, the master strides amongst the meditators, and when he sees a head drop, thwacks the delin-quent across the back with a flat stick – a *keisaku*. 'Pay attention,' he will bark – the injunction which is the foundation of all meditative and mystical practice.

Asceticism is rarely *only* about pummelling the rebellious flesh into submission. Generally it is about making oneself *available* to more than the demands of the flesh, and the most usual way of doing that is by cultivating more holistic ways of attending to whatever is both inside and outside. If one pays attention to more, one would expect to be able to see the connections between more.

The saints and the meditators, I expect, had to fall asleep so that they could reap the benefits of waking up, and so be for a moment in the place where many edges come together. The religions say with one voice that God is the ground of being. I've suggested that reality is a mesh of edges. You'd expect God to be particularly palpable where the mesh is palpably dense; and hypnagogia, say the hypnagogics, is one of those places. If God is reality, it's unsurprising, too, that other reliable insights into the nature of reality – whether about benzene rings or the behaviour of whirlpools – are given by hypnagogia.

It's not just the edge of sleep that's epiphanic. The fringes of other sorts of consciousness seem to be too.

I once had a splendid birthday present. I'd been moaning about distractions and restlessness.

'I'll sort you out,' said my friend Penny. 'It's your birthday next week, isn't it? I'm going to give you *yourself*. Come to our house at two p.m. on Tuesday.'

Not knowing at all what I was in for, and rather reluctant to meet with myself, I obediently cycled to her house.

'Lie down there,' said Penny. 'Watch my finger, and then count from a hundred down to ninety-six, seeing the numbers as exploding fireworks, with each number dimmer than the one before.'

I complied. By the time I hit ninety-six, my eyelids were fluttering uncontrollably.

'Now go into an absolutely safe place,' Penny told me. I did.

'From that place, wherever it is, there is a door. It leads to a garden. Go through the door.'

I did.

'In that garden there is a pool. It is wholly calm. It is undisturbed by the ripples of the past or the future. Look into it. You will see your face.'

With dread, I looked. I will not say what I saw there. But I will say that the technique of self-hypnosis, which Penny taught me that afternoon, was a great gift. I go every day into that garden, just over the edge of whatever is signified by ninety-six, and there is always balm there. I rarely have the nerve to look into the pool.

The greater gift, though, was one Penny didn't mention. I have discovered that if I snatch myself suddenly out of the garden, to a closed-eye place immediately beneath the surface of my usual consciousness, there is an explosive shuddering release from the strictures of thought, a leaping bonfire of vanities and presumptions, and a dazzling constellation of thoughts that seem new.

The great mystics – the thongs of whose sandals I am not worthy to untie – have, in their rosier times, a consciousness determined by its orientation with whatever they understand by the divine (and yes, I do have to put it in that convoluted way). But I notice, tremblingly, that the greatest work of the greatest of them is done when that orientation changes: when they go over the edge of that incandescent consciousness into the dark. Sometimes they return; sometimes they do not. Mother Teresa's letters show that for the last half-century of her life she was oppressed by a crushing sense of God's absence. 'If I ever become a saint,' she wrote in 1962, 'I will surely be one of "darkness". I will continually be absent from heaven – to light the light of those in darkness on earth.'[7]

But if they do return, and no doubt also if they do not, they are changed for ever.[8] They limp back out of the night, bearing terrible gifts that I devoutly hope never to see. Once more, a journey across an edge and a stay in the twilight do epic work.

Some have said it is *necessary* work. 'There can be no rebirth without a dark night of the soul,' wrote the Indian Sufi, Inayat Khan. 'A total annihilation of all that you believed in and thought that you were.'[9] I hope he's wrong.

There is perhaps a secular version of this (if anything is secular, which it is not). For it is often said that you can diagnose manic depression by looking at someone's CV. Manic depressives are, over their lifetime, hugely productive. They will have more publications, have spoken at more conferences, invented more widgets, won and lost more friends, catalysed more start-ups. The usual explanation for this phenomenon is the predictable one: the energy of the manic times (fuelled by the exhilaration of being sprung from the dungeon of depression) more than makes up for the languor of the low times. I doubt that's the whole story. It's more likely, I think, that creativity breeds in the seams between the highs and the lows; that novelty is gathered in the sickening transitions between the peaks and the troughs.

Whether or not this is right, perhaps we can agree with the writer at that dinner that ordinary consciousness is pretty dull, and a consummate under-achiever which ought humbly to give way to almost any other kind of consciousness. Ordinary consciousness is good for basic biological functions. Natural selection has presumably chosen and refined it for precisely those purposes – adjusting the input of information into our brain to exclude most of what is not absolutely necessary for feeding and breeding. This basic type of consciousness does not include subjectivity.[10] Nor does it include the type of consciousness engaged when writing or reading poems, listening to Renaissance polyphony or Taylor Swift, falling in love, wondering about meaning, feeling affectionate towards one's children in a way not adequately described by reciprocal altruism, kin selection or group selection, doing most of our ethical or other reasoning or, indeed, doing anything that makes life worthwhile.

I know lots of clever, creative mathematicians. They are, of course, very good at linear, logical reasoning. But that is not what makes them excellent mathematicians. Many of them – like the paradigm-trashing thinkers in all disciplines – describe the process of truly exciting mathematical research in something like these terms: 'I knew – I *intuited* – the answer. And then I spent the next ten years proving it.' The real discovery, in other words, happens over the edge of the type of consciousness on display in the mathematical journals. Their genius consists in their ability to commute shamanically between domains, grabbing insights as they go.

The same is true for our moral decision-making. It is not, for most of us, a matter of cold utilitarian calculation: the drawing up of a balance sheet with benefits on one side, detriments on the other, and a total score which, if positive, requires an action to be performed and, if negative, requires abstention. That tends to be how moral philosophy is done in the academy, where the conceit of philosophers and the privileging of a bloodless, bureaucratic, two-dimensional type of consciousness generates ethical formulae and hence hopelessly tidy ethical solutions, good for nothing in the real world. Real moral decision-making draws on a much bigger slice of the real world – diving deep over the edges of the calculating type of consciousness into ancient, resonant intuition. Intuition listens to the sophisticated demands of body and society and *mythos*, adjudicating wisely, not calculatingly, between them.[11]

Over the edge of conscious thought we are also over the edge of language, delivered from its rule over our thinking – free to flout the laws which, in business hours, might make some kind of sense but which fail to do justice to the complexity of human creatures, let alone to the baroque possibilities of our dance with the rest of the cosmos.

There are, of course, many other ways to leave our usual type of consciousness. The electric brain-storms of epilepsy have long been associated with radically different types of perception. In *The Idiot*, for instance, Dostoyevsky, an epileptic himself, describes the aftermath of a seizure experienced by Prince Myshkin as 'an ecstatic and prayerful fusion in the highest synthesis of life.'[12] The ways many will think of first are the ways we should think of last: drug-mediated alterations of consciousness, spontaneously altered states of consciousness of other kinds, and near-death experiences. Of the latter two I want to say little, except that they are enormously common across all cultures, and generally seem to be thoroughly good in their effects – often taking away the fear of death, convincing the subject that the universe is suffused with benevolence and meaning, and making subjects charitable and altruistic.[13] The same can be said for some drug-related experiences.

We see, too, examples of drug-related creativity and insight very similar to those seen in hypnagogia. LSD is said to have played a role in Kary Mullis' discovery of the polymerase chain reaction (the basis of the Covid PCR test), for which he won the Nobel Prize. It is rumoured (probably wrongly, in fact, but plausibly) that the structure of the DNA double helix was shown to Francis Crick under the influence of LSD. The psychologist and author Andy Mitchell, no stranger to chemical-borne travelling himself, comments that under the influence of psychedelics one's system of thought is 'temporarily unconstrained, allowing greater flexibility to imagine different possibilities. Thinking becomes hyper-associative.'[14] This is a now-familiar trope: our usual consciousness prescribes a certain way of doing things. Only when outside that consciousness completely can other possibilities be contemplated and realized. Our usual consciousness forbids access to anything other than a tiny slice of reality.

But it's not all good. Bad trips are common, and underreported in the typically blithe, rather romantic travellers' tales. They are very varied: psychonauts may have encounters with malevolent entities which, long after return to base, leave them with crippling flashbacks; they may be dissected by aliens, stalked by praying mantises the size of cranes, drink their own liquidized heart through a straw, or be gang-raped by swarms of newts. There may be good reasons for our consciousness's caution.

The general effect of psychedelic and other consciousness-altering substances is unsurprising. They use natural metabolic pathways – pathways used in sleep, ecstasy and alarm – but use them in unnatural ways. Sometimes that leads to an experience – good in its natural manifestation – being even better. Sometimes, as one would expect, misuse of the pathways produces a bad effect. Since we are naturally shamanic, air, properly breathed, is a psychedelic substance; sleep, if cunningly manipulated, or drumming, whirling or running up a hill can take you into the same zones as LSD or magic mushrooms.

It is no part of this book to preach about drug policy. My point here is simply to note that drugs, along with many other vessels, ply to and fro across the edges of our consciousness. Our dissatisfaction with the normal plane of consciousness – 'There must be more than this' – explains our insatiable appetite for anywhere other than here.

Perhaps this is just constitutional contrariness – a childish desire for whatever is (for good reason) forbidden. But the benefits and delights of edge-crossing endure when we come back to this plane of consciousness. It was useful to unravel the double helix. Salvador Dalí's dripping clocks resonate with the way we experience even nine-to-five-type time. Coleridge's poetry tells us more about the sorts of creatures we feel ourselves to be than does the Communist Party manifesto, or the doctrine of original sin, or our annual appraisal with HR. We seem to be made to inhabit a much bigger piece of reality than we usually permit ourselves to occupy. Another way of putting that is that we seem to be built, whether by natural selection, or God, or an alliance of the two, to walk along the edge of this plane, sometimes staying for prolonged periods on the other side.

Perhaps death is simply a more permanent relocation to the neighbourhoods we've already visited.

A 2017 study showed that human neural networks are wired to allow us to process eleven different dimensions.[15] We usually operate in just four of them: our three spatial dimensions and time. We can be so much more than we are. We can live in so much more territory. And perhaps we'll meet our beloved dead in dimension 9.

We're keen to see our dead again, but we don't want *personally* to cross the Jordan to see them on the Other Side. We're terribly attached to our places, our bodies and our comforts. And yet, in all mature traditions, and whenever we let down our oppressively cognitive guard, we hear an insistent, niggling denigration of *attachment*. It's almost as if, when our eyes are open, we know that we're designed for insecurity, and won't thrive without it.

Can that possibly be right?

8

Destitution and Death

. . . see the beauty in their lives. Hear the good news from the poor . . .

FROM 'BROKEN IMAGE',
song by Garth Hewitt[1]

For Malory [incarcerated in Newgate prison] writing was the last form of action available to him. As a prison writer, he needed the same attributes of endurance, defiance and survival that he represents in his imprisoned knights. Passive resistance in Morte d'Arthur *is anything but passive. Malory's knights construct a new definition of action, and hence of prowess. The duel, the arena of knightly identity, is transposed from the tournament to the dungeon.*

ROBERTA DAVIDSON,
'Prison and Knightly Identity in Sir Thomas Malory's
Morte D'Arthur'[2]

HERE IS A PROBABLY apocryphal story about St Francis. St Francis and his beloved friend, Brother Leo, returned from Perugia in the dead of a bitter winter night. It was so cold that icicles formed on Francis's habit, and kept striking his legs, drawing blood. They came to a Franciscan house and knocked on the gate. 'Go away', they were told. 'You are simple and stupid. We don't need you.'

Francis turned to Leo. 'Do you know what true joy is, Leo? I'll tell you. It is not curing the sick, or speaking many languages, or performing miracles, or knowing the deepest secrets of the natural or human world, or even seeing many souls snatched from the flames of hell by your preaching. It is this: arriving here shivering and wretched, and being insulted, turned away and even beaten, but accepting it patiently for the love of the crucified Christ.'

Is this an obscene story?

Like most serious questions, this can only be investigated by travel and immersion, and by sticking together little pieces of experience to form a collage, and then stepping back to see what's there. Any attempt at systematic philosophical or theological enquiry will founder and insult.

Here is part of the collage.

I drove out of Nairobi very early one January morning. When I'd broken out of the honking deadlock, I started to look around me. The road verges were alive with men and women walking. Many of the women had children strapped to their backs or traipsing along-side. They were all very poor. Their clothes were ripped and stained. I asked my Kenyan companion where they were going. They were going to look for work, he said. There was a place up ahead where, if they were lucky, they might be hired for some piece-work. If they weren't, they would keep on walking until they found something to do, or until they were too tired to carry on walking. But they were unlikely to get tired. They were very good at walking.

This was a desolate scene. I will not romanticize or sanitize it. I will only say that I was in an air-conditioned Land Cruiser, and not really in Africa at all. They had Africa under their bare feet; and would eat yam with their families that night, including with their dying parents, even if there wasn't much to go round, and I was four thousand miles from my family and wouldn't see them for weeks. When they woke that morning, whether it was under a flyover or in a lean-to in a makeshift city built of corrugated iron and asbestos on the edge of the city, they heard the mewing of kites and children and the cats that hadn't been stewed. I slept to the hum of a fan, and woke to CNN and the chink of coffee cups. When they got to the end of the road, past the industrial estates, they would hear the wind in the whistling thorn and sit for a while to watch the weaver birds before turning round for the long walk back.

I'm sure those people at the roadside would say that they'd prefer to be in the Land Cruiser or my cheerless hotel. But is that the end of the discussion about Francis's strange account of joy? Was he just a deviant?

We will see in Chapter 12 that Judaism has some robust and intricately choreographed taboos about contact with human corpses. A Cohen – a descendant of the priestly caste – must not touch a corpse. Others, if they do, must undertake cleansing rituals before being readmitted to the community of the living.

I have not been schooled in those taboos, but feel their force. I neurotically avert my eyes from hospitals. I'll take a big detour round hospices. Yet everyone tells me that these are the places I should go if I want to know how to live.

'I didn't know my mother,' said my friend Lucy, 'and she did not know herself, until the last week of her life. It was as if her cancer had eaten away not just her liver, but also her affectations. It had dissolved the shell that had kept her armoured against me and the world. It had ripped the veil that had hung for a lifetime in front of her eyes and stopped her seeing things clearly. Before her death, and *because* of it, she was remade: reborn. I don't know for what. It seemed a waste. Here she was, able to be herself for the first time ever, yet with only a few days to *be*. Sometimes it seemed awesomely kind that she should have been given this chance; sometimes it seemed sadistically cruel to dangle all that opportunity in front of her, only to snatch it away.'

This is a terribly common story. Sometimes it seems as if we're only fully embodied at the very edge of our embodiment – just as the embodiment is about to end. I'm reminded of Aristotle's belief that happiness (or rather his more holistic idea of *eudaemonia*), being a *goal*, can only be achieved at the very end of life.[3]

Lucy went on: 'She was *good*, too, in those last days. She never had been before. She was, to be honest, a rather snarky old woman, sniping and bitching and undermining. But all that side of her, too, was eroded along with her liver. She didn't become a saint, but she was warm and empathetic as she had never been. It wasn't that she was frightened of some sort of post-mortem judgment. She wasn't. Almost her last words were: "Soon I'll rot, and that'll be the end of me, and it'll be no bad thing." She behaved in a good way not for any purpose, but simply because she *was* good, somehow.'

This, again, rings Aristotelian bells. The good of humans, wrote Aristotle in the *Nicomachean Ethics*, is good functioning – which is the activity of the soul in accordance with virtue.[4]

At the edge of biological life, Lucy's mother both saw things more clearly (as one might expect, for edges, as we've seen, are the only places for really good views) and was also, less intuitively, more virtuous.[5]

That night in Nairobi, in another anonymous hotel giving round-the-clock service to commercial travellers, Kenyan prostitutes, and shareholders who wouldn't dream of leaving Stuttgart, I sat in the bar, opened a notebook, and tried to sort things out.

Humans are moral, I wrote – and I believed and believe it. Morally good, morally bad or, usually, a mixture of both. But incurably moral.

The beer was bad, and there was too much of it. The notes were incoherent. But, scrubbed up, this is the gist: morality propels and paralyses us. It inhibits our worst and sometimes our best, cheers on our best and sometimes our worst, coheres and divides our families and societies, is the tool of political charlatans and the bread of saints. Morality feeds the hungry and, less often, starves the fat. Some say it streams down from God; others that it oozes up from our evolutionary past. Some say its nature is law; others that it is really grace or instinct. Both its presence and absence can cause great cruelty. It is stern and tender, easily overridden but generally robust under trampling hoofs. It may have slightly different colours in different places and times, but its consistency is far more remarkable than its fickleness.

Then I went to bed. I didn't sleep. I got up and stared through the double glazing towards the road I'd driven down that morning, and then sat on the bed turning platitudes round and round in my head.

That sort of night rarely helps. But travelling does.

Every traveller knows that morality (like everything) is best seen at the margins.

Once, for instance, walking stupidly and irresponsibly along a camel-herders' route in Sudan, I miscalculated everything: times, distances, water. Hallucinating with fatigue and dehydration I saw, a mile off, a cluster of black tents. That, at least, was not a hallucination. I stumbled into the camp. Children stared at me, but Mohammed, the grandfather, *saw*. He beckoned me into a tent, took my rucksack from my back, gestured to a pile of cushions and forced a goatskin of water into my hands. I was asleep in ten minutes. When I woke it was dark and there was a smell of cooking. They had killed a kid for me, and we sat into the night, tearing off gobbets (I had to go first with each round of eating) and smiling solemnly and silently at one another by firelight.

I stayed, guiltily, for two more nights. They wouldn't let me go. They had nothing, and gave everything. I tried to make amends: to give them something. I thought they might at least take my Swiss Army knife. They wouldn't. No, they said, my visit was enough. I had honoured them. At last they took me to the road, hugged me, bowed, and moved back into the desert.

These ancient Bedouin mores may have their evolutionary origins in reciprocal altruism: we give hospitality because one day we might need it. I scratch your back so that you'll scratch mine. But the origins were irrelevant to the holiness of their gift to me. I could not scratch their backs. I would never see them again. If it was a transaction with roots in any shared fellowship, it was the fellowship of the edges.

The kindness of strangers is one of the foundational anthropological facts. It is also one of the strangest. And it is dramatically on show at the margins. If I'd turned up at a big house in Khartoum they'd have set the dogs on me. Only from nothing comes everything.

Isn't this true, too, in areas of morality a long way from hospitality? For forgiveness, for instance – one of the rarest, most arduous and most vital disciplines. The most hurt, the most dispossessed, the most horrifically bereaved are often those who hate least and forgive most abundantly. The mothers of dead soldiers are in the vanguard of the peacemakers, readiest to embrace the enemy. It is the least affected who call for vengeance; for the flattening of towns, for two eyes for every eye and ten teeth for every tooth.

I'm not romantic about poverty. I'm certainly not romantic about morality. We see bad as well as good morality most clearly at the edges. Edges are fertile ground for everything, including evil. Go to what you'd have thought were the limits of human depravity and you'll find more ingenious and energetic and creative depravity springing up. For every decency forgotten and every restraint removed, a thousand demons are remembered and a thousand fiendish liberties enacted. Kurtz is real, and his address is the Inner Station (a name that hints that he has created his own centre) beyond the Pale.[6] The edges, then, are where this defining human quality – morality – is most on show, both in its presence and absence.

The theologian David Bentley Hart writes that the first generation of Christian believers were a 'company of radicals' whose values were '... almost absolutely inverse to the recognised social, political, economic, and religious truths not only of their own age, but of almost every age of human culture.' They 'bore very little resemblance to the faithful of our day, or to any generation of Christians that has felt quite at home in the world, securely sheltered within the available social stations of its time, complacently comfortable with material possessions and national loyalties and civic conventions.' Most of us, he declares, 'would find Christians truly cast in the New Testament mold fairly obnoxious: civilly reprobate, ideologically unsound, economically destructive, politically irresponsible, socially discreditable, and really just a bit indecent.'[7]

This rag-tag tradition of disreputable, unwashed, patched, anarchic, dog-on-string-leading, tin-whistle-twiddling, insolent, merry, free, solemn, puckish, laugh-and-weep-till-they're-sick-at-the-sight-of-your-big-car, idiotically generous prophetic iconoclasts has never entirely vanished, though it took some bad blows from Byzantium, and some worse from Rome, and some worse still in the grim citadels of the Reformation and the megachurches of the US Bible Belt. But Holy Fools, mooning at the altar, have an honourable place in Russian Orthodoxy. Francis, the holiest and most foolish of all since Jesus, has never quite been silenced by all the reverence poured on him; has never quite been constrained by the gigantic church built to honour his cult of destitution.

Wherever morality comes from, it doesn't come from the centre – whether from a parliament, a metropolis, an overarching theory, a canonical idea, or an episcopal palace. The former Bishop of Edinburgh, Richard Holloway, observed: 'As a matter of fact, moral change always comes from those who are out ahead of the institution, never from those in charge of it.'[8]

Is there a better word than 'morality' for what I experienced in the Sudanese desert? I think there is. But it is very unfashionable. 'There is a word so ancient that you can hardly say it in sneering metropolitan circles,' writes Jay Griffiths. 'It is *honour*.'[9] Its absence from the metropolis is the most troubling political fact of our age. Honour describes both proper comportment and the reason for that comportment. It is kin to dignity – one of the attributes of the gods, one of the aspirations of Stoicism, and, according to a maxim of astonishing *chutzpah* and deafening resonance, held in the Judaeo-Christian tradition to be an inalienable characteristic of all humans. Good living, in that tradition, is living in accordance with one's dignified nature. We saw it in Aristotle. We saw it in the hospice where Lucy's mother was dying of liver cancer. We saw it on that Kenyan road.

'Death destroys a man,' wrote E. M. Forster, 'but the idea of death saves him . . . Squalor and tragedy can beckon to all that is great in us, and strengthen the wings of love. They can beckon; it is not certain that they will, for they are not love's servants. But they can beckon, and the knowledge of this incredible truth comforted . . . '[10]

Honourable and dignified conduct, like everything else, improves with practice and exercise. For humans, human existence is the gym designated for the training. The more strenuous the gym, the faster and fitter your honour and dignity will grow. And where are the most strenuous gyms? At the edges, of course. At the centres, all the really effective gyms seem to have shut down. That's dangerous.

Some of the surviving gyms are in churches, mosques, temples, synagogues, meditation halls; dusty, under-visited corners of the mind, sweat shops, soup kitchens, roadsides, hospitals, landfill sites and cemeteries. And the most helpful exercises for our time in the gym are all – if we could only see them properly – steps in some sort of dance; a dance in which we're only one dancer in a great multitude. Most of our fellow dancers are gloriously dead.

Some of the best practice and exercise is done in theatres, cinemas, libraries and galleries. We'll go there next.

But Francis? What about him? A pervert? Grievously mistaken? Wrongly translating to this life an edict that relates only to eternity?

Is there a moral obligation to abolish poverty? Absolutely. Unequivocally.

Might that be an obligation to abolish the greatest joy?

Who knows?

Those who have least give most. This is an immutable law. And, some say, it is more blessed to give than to receive.

I can say no more, either for Francis or against him.

9

Dogs on Cornflake Packets

A good writer always works at the impossible. There is another kind who pulls in his horizons, drops his mind as one lowers rifle sights. And giving up the impossible he gives up writing.

JOHN STEINBECK,
Journal of a Novel: The East of Eden Letters[1]

THERE WAS A GENTLE hubbub and a clinking of glasses of warm white wine as I pushed through the throng at the gallery somewhere off Bond Street. My friend Terry was at the back. Unlike anyone else he was looking at the exhibits we'd been invited to see. I could see the sneer through the back of his head, and braced myself for the speech I'd heard so often. He was true to form. I didn't bother to follow the argument, but just ticked off the familiar words as they tripped off his well-lubricated tongue. 'Childish'; 'No, infantile.' 'He's having a laugh.' 'Do they know they're being mocked and defrauded?' 'Any reasonably coordinated five-year-old would be ashamed, and its parents in despair.' 'What do these people think when they see *real* art, d'you suppose? Do they really think this has anything in common with the ceiling of the Sistine Chapel?'

And so on. I didn't have to follow closely because I'd taught Terry most of the lines. Now, though, I was wondering if I should have done. True: you didn't need to go to art school to pile those cans on top of one another, or to compose the title: *Pile of Cans*. ('Not so much a *Pile of Cans*, said Terry, as a pile of . . . ' Yes, yes, Terry, very amusing.) True, you would need an urgent psychiatric examination if you paid the eye-watering price demanded for the pile. True, there were far too many adjectives in the gushing catalogue description, and no doubt the artist was sniggering all the way to the bank, having read that the teasing cleft between the tomato soup and the meatball tins made him the next Klimt. If Tate Modern bought it, it would have to keep it on display and plead its case for years, deploying a new set of adjectives simply and solely because it had paid so much for it. Yes, comparisons with Michelangelo or Klimt were absurd and an affront to common sense, let alone scholarship. But I wasn't as ready as once I was to mock, for I was slowly beginning to understand that if I did, artists such as these would have the last laugh.

'Whatever it is, it's not *art*, is it?' Terry was saying.

Wasn't it?

It had, after all, made me ask 'What is art?', which was itself an artistic achievement. It had forced me into a conversation with the artist, which is what it set out to do. No matter that the conversation was rancorous on my part. It had made me wonder whether my kids, all of whom could do better than that, were artists – and I went back to them with renewed respect, looking more closely at their daubings. *Pile of Cans* was a scaffold from which, if I climbed to the top, I might see something new; and another kind of scaffold on which I might execute some of my preconceptions.

The artist saw us looking. He came over, stood beside us and asked us what we thought. He was a pale young man in a linen suit too big for him. He needed a good feed.

'Interesting,' said Terry. The word that can stand for the highest praise and the foulest calumny.

'Well, thank you,' said the young man, who didn't appreciate the ambiguity. 'I hope, anyway, it helps you move on.'

And he moved on himself to the next knot of people, acting on his agent's instructions to press as much flesh as possible.

Move on. Move on?

What on earth did he mean?

Terry was in no doubt. It had indeed helped him to move

on – out of this den of pretension and to the pub, and was I coming? I was indeed. But the question stayed with me, undiluted by the pints that followed, and as I trundled back to Oxford on the late train, I began to see what the artist meant.

Art is what you do on blank pieces of paper, blank cave walls, on plinths in Bond Street which have nothing else on them, or with lumps of marble you think have a god trapped inside. It's about populating emptiness with something of you, rather like a reproducing amoeba splitting its body in two. It's about snipping off something of yourself that then crosses the boundary that usually separates the you and the not-you, and which can be seen and touched by others who aren't you, and start a conversation. Even if it's a contemptuous conversation, you'll feel less alone for having had it. Once it's on the paper or the plinth or in the stone it won't die, or at least not when you do. It's a sort of insurance against death.

Isn't this why we do *everything* we do: to feel that we are not alone, and to feel that we will not die? If we can cross our own edges – out of our mortal bodies – there's a chance of salvation. A state of perfect self-possession is a state of wretched loneliness, and worse, it's a state of the doomed, because self-possessed bodies will die and rot, along with the Self. Get out of ourselves: only outside ourselves do we have any chance. Hence our curiosity about everything: our constant staring over borders; our obsession with *otherness*. Why else would we read? Or talk? Or have animal pets we call our 'fur babies'? In *Crimes and Misdemeanors* (1989), Woody Allen declared that ageing men chase younger women because they fear death: meld with another body and something of you might survive. Reading groups, social media, doggie coats and the desperate adultery of the senile come from the same place inside us.

Move on; move out; the status quo, which is the state of self-satisfaction, is out to get you.

Move on; always on.

*

A toddler was scribbling (at least that is what it looked like) on the back of a cereal packet. She put down the crayon, tottered back to examine her handiwork, and handed the packet to her mother.

'Lovely, darling. What is it?'

'A dog.'

'Of course. A wonderful dog.'

The toddler was engaged in *the* quintessential human activity. She was making a new thing. When we're said to be made in the image of God, alchemy is surely one of the main characteristics of the image. God spoke, and from His word something that had not existed sprang into being. Toddlers scribble, and a dog, which had not existed before, springs from the back of a cornflake packet. The cornflake packet had moved on and become a dog. The toddler had moved on and become a dog-mother. Sometimes human words effect the act of creation, making new stories or new connections or, more modestly, turning an existing entity around so that new light shines on it, revealing a side previously invisible.

A dog had come out of the toddler, crossing the edge which delineated the toddler's boundary. The toddler had created a new set of edges. That's how edge-people reproduce: by multiplying edges. And, importantly, the dog had been shared. Normal people show their art to others. Art is relational – and in this sense too it is about transcending the edges of the Self. We can't help birthing, and we can't help showing off the baby. We long for the 'Ooo, isn't she beautiful!', and both from the act of reproduction and the exclamation we are less alone. The toddler was learning how to mitigate loneliness.

The toddler's dog was certainly art. It did, and was at least partly intended to do, exactly what the *Pile of Cans* and the ceiling of the Sistine Chapel do: push out a foetus that just has to come; increase the number of entities with whom relationship is possible; elicit a response; boost the chance of immortality; make everything move on; transcend edges and create new ones.

*

There's a tendency to think that art as boundary-transducing, paradigm-smashing, transgressive and *edgy* is a feature of modernity, if not modernism, if not postmodernism; a reflection of a recent restlessness, created because we've become unmoored from our metaphysical anchorings. Here is the German literary scholar Claudia Oik:

> The history of modernity is a record of inherent restlessness. Modernity is not a stable condition established in perpetuity. It is always on the edges, verges towards historical, discursive and aesthetic thresholds and reaches beyond them . . . The discourse of modernity is by necessity incomplete and in need of constant revision and readjustment.

She quotes Neil MacGregor's observation about 'Shakespeare's restless world', and continues:

> [F]rom the sixteenth century onwards the sense of experiment, the transcending of limits and emphatic novelty have been conceived as intrinsic to the idea of a modern age that does not aspire towards an unchanging status quo but implies its own continuous demise in an inbuilt dynamic of self-supersession.[2]

This is a splendid description of modernity. It is also a splendid description of every time that has ever been. Every time was modern to the people living then. The restlessness Oik describes is intrinsic to the human condition. The 'continuous demise' that fuels the 'dynamic of self-supersession' is not the demise of an age but the demise of individual humans and their confidence in their ideas. It comes from the recognition, however tentative and under-processed, of the transience of our bodies and our notions.

An 'age' (whatever that is), a political system or a nation might aspire towards an unchanging status quo, but every human knows that, for him or her, no status quo is possible; that very, very soon they will be eaten by worms. And so we pick up our paintbrushes and tell our stories to deal with it; to cross the boundaries of ourselves and our defining conventions, because inside the boundary is the stuff that will rot. Just as humans can't do anything that isn't religious, we can't do anything that isn't at least potentially art. And art is by nature transgressive. The most important transgression is of oneself: it's about putting something *out there*: it's the business of propulsion of something from oneself.

A. N. Whitehead put it like this: 'Great art' (not a term I like or understand) is something which 'transforms the soul into the permanent realization of values extending beyond its former self.'[3] The Self, by producing art, moves out and on: the former self is left behind (hurrah, for it is doomed) and something permanent is realized (hurrah, something of us will survive). 'Any experience carried out deeply to its ultimate leads you beyond yourself into a larger relation to the experience of others,' Anaïs Nin told her diary.[4]

We can't stop being artists, any more than we can stop breathing. Ever since we've been *us* – behaviourally modern humans – we've been artists. We've seen that already. Ice Age art makes Picasso at his most anarchic look like a car salesman at home in Surbiton, dozing in front of the TV. Picasso, to give him credit, knew it. If those Upper Palaeolithic hunter–gatherers were the quintessential humans, we'd expect them to produce the quintessential art. They did.

'Art,' said the German Romantic painter Caspar David Friedrich, 'is about the crossing of thresholds.'[5] Quite right. And so if there is such a thing as a great artist, it is presumably one who moves across a greater number of edges. Michelangelo is said to have created a new style of art every time he picked up a brush or a chisel – to the despair of his would-be competitors. Yet surely he was different not in kind, but only in degree, from the others. If the metric is the number of edges crossed, I suspect that the toddler with her cornflake packet would outrank the great Florentine master.

You'll cross more edges the greater your motivation to get out of yourself and far from yourself. That, perhaps, was the key to the genius of broken-nosed Michelangelo. 'I am ugly,' he said, as, in *David*, he created the matchless archetype of male beauty. *David* was as far from Michelangelo's self-image as a human figure could get. Lucian Freud, a supreme painter of human vulnerability – of our unstable pose on the edges – was displaced from Berlin and from himself, and spent a life loathing himself, running from himself to the bars of Soho and back to his easel.

Just as those early shamans may have given us the very capacity to taste art, as well as the content of some of the best art there has ever been, the very business of art, which is the business of shuttling across divides, is shamanic: it involves going out of yourself, picking up something from another world, bringing it back, and delivering it to someone other than yourself. 'Shamans have traditionally lived on the edge of their communities,' notes Jay Griffiths, 'and the quality of "edge" is what marks original artists.'[6]

To be a regular and efficient shuttler between continents, it helps to live near Heathrow. The shamans, as we saw with my neighbour Mrs Bogg, live on the edge of their communities, for there is no very reliable distinction between metaphorical and actual edges – or indeed between the metaphorical and the actual. It helps a painter or poet to starve in a garret, as long as it isn't an artistic pose.

Once I starved in a winter wood for a while. The pain, followed by resignation and weakness, removed my ability to be conventionally coherent. I found it hard to look directly at a tree or an idea. But it did wonders for my peripheral vision and the margins of my other senses. I smelt my dead father's pipe smoke and corduroy; I knew where the hare would come the next night; I saw ghosts in the hawthorn. I don't doubt they are there all the time.

If I turn my head quickly in the heat of the day in southern Greece, I may see the shadow of a faun or a hamadryad. They are as invisible to direct sight as the feeling of a poem or the meaning of a piano concerto. Perhaps nothing worth having can be seen in plain sight, for the same reason that nothing worth having can be measured, weighed or accurately described. Real worth can be seen only out of the corners of our eyes, because it is in the margins of the cosmos itself. C. S. Lewis observed: 'It looks as if the Romances and such Ballads were in the Middle Ages, as they have remained ever since, truancies, refreshments, things that can live only on the margin of the mind, things whose very charm depends on their not being "of the centre".'[7]

There in the margins, everyone is a Michelangelo.

Much artistic inspiration comes from a place unknown to the artist. Virginia Woolf's description of the genesis of *To the Lighthouse* is typical.

> . . . one day walking around Tavistock Square, I made up, as I sometimes make up my books, *To the Lighthouse*, in a great, apparently involuntary, rush. One thing burst into another. Blowing bubbles out of a pipe gives the feeling of the rapid crowd of ideas and scenes which blew out of my mind, so that my lips seemed syllabling of their own accord as I walked. What blew the bubbles? Why then? I have no notion.[8]

It's a description of something rather like possession or clair-voyant channelling. The feeling is, anyway: where on earth did *that* come from? The writer and anthropologist Eric Wargo, in *From Nowhere*, contends, buttressing his thesis with hundreds of examples, that the *usual* way in which artists get their big ideas is by eaves-dropping on the future.[9] Being an artist therefore involves having a finely developed precognitive faculty. The inspired idea tunnels back to the artist from the future, popping out into their consciousness through one of the quantum physicists' wormholes in space–time. No wonder artists are surprised.

If Wargo is right, we've identified yet another edge crucial for human culture: the edge separating the present from the future.

*

Humans, being edge-organisms, desperate for solidarity with other edge-organisms, are moved and entertained only by dances on edges or moves across edges. Things are only funny if they are at the edge of taste, comfort or probability – and funnier the more they resonate with our own sense of edginess. The funnier the joke, the more impossible it is to explain it, because language (a very staid, centrist medium in day-to-day use) cannot reach that far over the edge. Monty Python's 'Fish-Slapping Dance' is funny because . . . well, why? But it is. It just is.

George and Weedon Grossmith's magnificent *Diary of a Nobody*,[10] which tells the non-story of the non-doings of a clerk, Charles Pooter, of The Laurels, Brickfield Terrace, Holloway, works precisely and only because it is a satire. A deadpan account of an uneventful, edge-free life in Holloway wouldn't work on any level. Boundaries have to be crossed in order to make us read, watch or listen. Staying at home in your slippers doesn't make for good art or good living.

In John Buchan's *John Macnab*, three pillars of the Establishment are suffering from deadly ennui.[11] Their remedy? Leave the metropolis and head to the Highlands of Scotland to do something criminal – in their case poaching a stag or a salmon. It's a better story – and the healing far faster – than if they'd stayed with stiff upper lips at their desks.

Sir Gawain and the Green Knight opens in the centre of centres and the centre of comfort – Camelot at Christmas, 'amid merriment unmatched and mirth without care . . . din of voices by day, and dancing by night.'[12] At New Year '. . . double dainties on the dais were served . . . ' But Arthur is restless – nauseated by the comfort, dismissive of the 'high tables of trifles courtly.' He knows that this is not how humans are meant to be. We're built for adventure. The hunter–gatherer instinct stirs in him. '. . . [I]t pleased him not to eat upon festival so fair, ere he first were apprised of some strange story or stirring adventure or some moving marvel that he might believe of noble men, knighthood or new adventures . . . '[13] His royal duties mean that his adventure will have to be vicarious. Gawain wanders for him. It means wilderness; crossing the boundary between humanity and the non-human world, and many other boundaries too. From the start he has 'no friend but his horse', 'no man on his march to commune with but God ... '[14] He reaches the Wilderness of Wirral, where there are 'but few who with goodwill regarded either God or mortal', taking 'pathways strange by many a lonesome lea ... Many a cliff he climbed o'er in countries unknown, far fled from his friends without fellowship he rode.'[15] There could not be a greater contrast with the warmth and silkiness of Camelot.

Gillian Rudd has pointed out that in the Wirral everything is so strange that names disappear: there are instead 'generic cliffs, fells, naked rocks, and water ... '[16] Gawain has none of the certainties – even about the identity of things – that are the stock-in-trade of the centre. The inhabitants are hostile: snakes or dragons, wolves, bulls, bears, boars and 'wodwos' – wildmen, who may be feral humans, or perhaps another species altogether. As he plods on, even these creatures fall out of the story. His own horse is not acknowledged. He's over the edge of the known world.

Where does the final action happen? In another courtly centre, with fluttering banners and thrones and trumpets? No: in the Green Chapel, which is 'nobbut a cave'. The ultimate non-centre. But Gawain does visit a castle before the denouement. There is a reprise of the scene at Camelot. At dinner, 'marvellous dishes were set on the dais, the daintiest meats'. Gawain and 'the gay lady together were seated *in the centre*'.[17] This return to the centre, of course, threatens the whole project.

Diary of a Nobody, John MacNab and *Gawain* are all very good stories. All very good stories, whether they are set in Holloway, the Scottish Highlands, or a wilderness with no names, are the same story.

Overcome the centripetal drag if you want to be what you are. It's the consistent cry of the ages and the consistent cry of any writer worth reading or painter worth paying to see. Away with the delusion of certainty. 'Art *destabilizes*', writes James Cahill. 'Art is the same thing as life – and life is dangerous and uncertain.'[18] Out, out, if you want to achieve anything in life. Here is Thucydides, eavesdropping on a Corinthian description of the Athenians: '... while you are hanging back, they never hesitate; while you stay at home, they are always abroad; for they think that the further they go the more they will get, while you think that any movement may endanger what you have already.'[19] It might be advice in a masterclass for stand-up comedy, or novel-writing, or landscape painting. Even the apparent glories of the centre depend on *leaving* the centre, as Pericles recognized in his oration at the end of the first year of the Peloponnesian War: '... our adventurous spirit has forced an entry into every sea and into every land ... You should fix your eyes every day on the greatness of Athens as she really is, and should fall in love with her. When you realise her greatness, then reflect that what made her great was men with a spirit of adventure ... '[20]

Real stuff, whether it is the 'Fish-Slapping Dance', a Tallis motet, a ham sandwich, the love of a child or the appeal of a mountain, can only be tasted, not described. Reality, being at its root relational, can only be encountered directly: one on one, I to Thou. Since our vision and cognition are catastrophically entangled, to see something directly is to translate it into our own thoughts about it, and so turn it into a reflection of ourselves. I want to relate to that mountain, not my own thoughts about it. Hence the need to see out of the corners of our eyes, and to process with the corners of our minds.

But some things sometimes have to be said. Poetry is not always worthless, or incurably self-referential or self-reverential. How can words be worthy?

Except when they are brutally utile – such as when we ask someone to pass the salt – they must, if they are to relate honestly and meaningfully to the world, be metaphorical. Metaphor is what makes us *us* as a species. It was the basis of the cognitive revolution that made us what we are. It is the core of that defining, incontinent, incessant, obsessive symbolizing: behind the making of lion-headed men from mammoth ivory and the imprinting of red ochre hands on walls deep inside the earth.

Metaphors have wings and lives of their own, like everything else that is real and distinct from us. We need mediators to cross the edges between us and other sectors of reality, and metaphors are the ordained mediators.

Consider a metaphor: *the cat is a stone.*

It does tectonic work.

It asserts the unity of very different things.

It makes us laugh.

It makes us wonder about the hinterland, and perhaps the sanity, of whoever coined it.

It interrogates and explodes our presumptions about cats. And about stones. It reshapes a small corner of reality.

And that's before we start on the positive work, demanded by its author, of inquiring into the ways in which cats and stones might indeed relate to one another. That process entails a very explicit and robust understanding of relationality itself. We step out of ourselves and into the big jungle where cats, stones and everything else are entangled.

The metaphor harnesses the particularity of cats – their individual delimitations – to tell us something about the particularity of stones. And vice versa. And both, separately and together, tell us general things about the way things are in the world. Have you tried to pick up a sleeping cat? It's hard, because the cat (being a stone) is much heavier than itself. Stones (being cats) couldn't care less about us, and survive even if we don't feed them. Whoever thought that cats and stones could be so fecund? The metaphor shows us that they are, and if *they* are, what about the whole of reality!

If we inhabit the world of metaphor (and what other world is there for us?), cats are stones, while still remaining cats. This is an exciting place, as we discovered about 45,000 years ago, when we noticed that a bit of mammoth ivory could be a lion–human hybrid and a dead father while still being a bit of mammoth ivory.

If cats are stones, then cats are potentially stars, oven gloves, sunflowers, and indeed everything else. This is what Vedānta (the branch of traditional Indian philosophy based on the Upanishads) has always said – though with a slightly different accent. Our metaphors link everything with everything: they draw connecting lines. The metaphorical world is densely criss-crossed with lines. If everything's linked, to go *anywhere* at all outside yourself is to go *everywhere*: to cross all the edges there are.

To say that we are metaphorizing creatures is grand, but not grandiose. It is to say only that we speak; that we think in language; that we create mental and physical pictures. But some language and some pictures are worthy of our status as metaphorizing animals, and some are not. We shouldn't conclude that something is worthy of us because it looks or sounds or smells arty. Artiness is often the sworn enemy of metaphor.

Yet metaphor is *the* artistic project, and when we use lines, whether on a canvas, or in a written sketch of character, or in an argument, those lines are powerful – and so worthy of us – only insofar as they hint at the mesh of connecting and dividing lines which make up reality. Metaphor, as those mammoth-ivory whittlers knew, is about knowing that everything can be everything else: that lines can be drawn between everything. It is our instinctual affiliation with metaphor that fuels our obsession with more ordinary lines.

The obsession is on show everywhere. Look at the couples, hand in hand, walking along the seam between the sea and the land; at the student philosopher earnestly making distinctions between categories; at the taxonomist, labouring for a lifetime to make the case for moth A being different from moth B, and at her evolutionary biologist colleague in the lab next door, showing that not only all moths are basically the same, but that *Tyrannosaurus* and mosses are basically the same; at the house-proud husband, unable to sleep because the cupboard and the sink aren't properly aligned; at the serried ranks of named pigeonholes; at the tidy propositions in the Nicene Creed; at the journalist tracing the connection between the politician and the drug baron. Listen to the clocks and fireworks announcing the turning of the year, and to the chant of the bigot as he tells the foreigner to get back to where he came from.

Some of these examples assert rather than deny boundaries, but both the assertion and the denial affirm the importance of edges. Each example might be the core of a workmanlike story, but you would never think the story worth reading unless the edge involved were crossed.

We get to the general through the particular; never the other way round. Particularity means edges. That was behind Blake's stridency on the subject of lines: 'The great and golden rule of art, as well as life, is this: the more distinct, sharp and wiry the boundary then the more perfect the work of art . . . '[21] His works of art all sought to say something universal.

Though he said precisely the opposite, Vermeer was making the same point as Blake. He loathed lines, which is why he never drew any – believing there to be no lines in the world. This is a vibrant statement about connectedness, but if we conceive of connection as that infinite number of connecting lines (I'm sure Vermeer would approve), Vermeer and Blake end up in the same place. Blake wants to honour oneness through particularity; Vermeer to honour particularity (which is what he painted) through oneness.

In this old stone house on the edge of Europe, where I live for some of the year on the edge of hygiene and decency, I have a self-denying ordinance which, like many such ordinances, is really sybaritic. When I'm there by myself, which is most of the time, I allow myself electricity only to power the old laptop on which this book is written. The light comes from the sun, from an olive-wood fire in the corner nearest the sea, from the swinging light in front of the icons, and from old-fashioned oil lamps with tall glass stems.

It sounds ridiculous. Pretentious. Why do I do it? Luddism? No, though I've a lot of sympathy for that. A romantic self-image? Absolutely not. Simply because the flickering casts more shadows than the one set of shadows cast by the electric light. I get more edges. The room is deeper and odder. The shifting shadows expound it. Electric light flattens it; appeals to my visual reductionism; tells me that there is one definitive way that the room is. I get more information from the flames; I squeeze more living out of the room and my time in it.

But I must be careful when talking about 'points' or artistic theories or messages. Nothing worth reading or going to a gallery to see is about a *message*. Messages are centre-things, invisible in the margins where everything happens. Salvador Dalí declared that he had no message. I had doubts about his authenticity until I read that. Encounters across the edge of Self have no message. Encounters just *are*. Dalí said too that he didn't like steak because it had no face. Encounters are more intense if the other has a face.

It's better just to say that the oil lamps help me to be in that room better. Or perhaps that they help me to see faces in the wall.

For me to denounce all messages and declare that Michelangelo and the *Pile of Cans* and the cornflake-packet-scribbling toddler are about the same thing risks an important misunderstanding. It is a great error to confuse centrism with order, and ediness with anarchy.

Squeezed between a mountain and the sea is a village square. Presiding over it is a green bronze statue of a fierce man, his belt stuffed with daggers as sharp as the ends of his moustache. In front of him, real children are dancing. The boys have black stockings on their spindly legs, and boots their grandfathers died in. The girls wear headscarves and necklaces made from the coins that closed their grandmothers' eyes. Their hands are on their hips or on the shoulder of the next dancer. The girls' skirts nearly touch the ground, but you can just see their feet. Every foot in the line is doing the same as every other. The music is being scraped with a bow out of a lyra. There's a sense in which it is Homeric; another in which it's brand new, because the player has never done *quite* that before; and another in which it's as old as the chemistry of the mountain and the swirl of the sea and the genealogy of the pig whose buttocks contributed to the kebab that's powering the player's hands.

One might say the same about the play of the feet, but it would take a far keener eye than mine to see the difference between the inflection of *her* right foot and *his*. The movement of the feet is as intricate and as fast as the slides between the quarter-tones in the music, and yet the lyra-player's fingers are so fast they're blurred and it looks as if he's lost half his left hand to frostbite.

Those feet are following a very old script. When the children were practising, the teacher roared at them when they departed from it. It would be heresy not to touch the right heel there and the left toe there. Heresy is policed seriously in this kind of country. Perhaps it looks as if the feet are falling in with a decree from a centre some-where. They're certainly travelling, with every heel-toe and twisted instep, along a road well travelled. They're orthodox.

But where do the commands come from? From the dead: over the biggest edge of all. The tunes, if they weren't transcribed by Odysseus after he'd heard Athena singing them, or spun from spume and the cough of ravens, were croaked out by long-dead farmers after the third jug of wine. They are often laments; speeded-up versions of the wailing you hear at funerals in these parts. Or about lost love or lost goats – but in any event about the losses suffered by those who are themselves lost. And the steps themselves: they too are the property of the dead; it's the dead, not the teacher, let alone an institute in Athens for the maintenance of Greek dance, who are wagging their fingers when a heel goes down instead of a toe.

Now that I watch the dance more carefully I see loss there too: the lad who is waving his handkerchief and moving down the line, flirting with one girl, leaving her, promising and acting out consum-mation with another, leaping high like a ram to show that his semen is best, is announcing that everyone else in the line is bereft and will die alone with their cats in a cottage hanging from a cliff. This mer-riment in the sun, caught on the phones of the adoring parents, is orchestrated by the dead, for the dying, and its subject is the passage over the edges that run through life.

If the feet weren't obedient and disciplined – if they weren't orthodox – none of this edginess could be there. It'd be a random thrash, signifying nothing, encoding none of the sultry, explosive *eros* and *thanatos* of this dance. And there is, too, the irony that is no irony: that service is perfect freedom. That girl there is far more herself because of her obedience. That immaculately choreographed tilt of her head is far more hers because it was done in clairvoyant connection with the community of the living and the dead in whom she lives and moves and has her being. By moving deferentially over and between the edges mapped by the journeys of others' feet and necks and marriages and funerals, she affirms herself as an edge-person. And if you want to be a person, that's really the only kind of person you can be.

We've been on a tour from the Big Bang to the nation state, via the evolution of sex, the shapeshifting of shamans, the spearing of gazelles, the genius of drunkenness, the death and rebirth of gods, and the aristocracy of the destitute. We've stopped off in swirling galaxies, islands penned behind walls of green water, cemeteries, mosques and private views at central London galleries. We've sat and slept beside great, guilty rivers, visited the first European in a Peloponnesian cave, nearly died from stupidity in Sudan, felt guilty and got all introspective in Kenya, failed to construct a coherent political philosophy in Thailand, drunk retsina with dying Greek sailors, and savoured impostor syndrome, washed down with port, in Oxford. We've toasted *Gawain*, denounced Chaucer, and expressed scepticism about the very language in which this book is written. There's been plenty of assertion and precious little argument. We've looked at lots of particulars.

Now it is time to look back and look harder at some of the things we've seen, and wonder together – while remaining deeply suspicious of abstraction, and avoiding bullet points as the work of the devil – whether what we've seen might fit into some tentative pattern. We've seen some edges, but what is *edginess*, and how does it behave, and why? What does living on edges, in a world made of edges, mean for our nature as humans? Why do we recognize edges, and think them significant, and mark them?

PART 3

Edginess

10

What Are We? Where Are We?

Warning: There is a bit of very gentle philosophy here.
If that's not your thing, skip to Chapter 11

Petrarch climbed Mount Veritoux. He took out a copy of Augustine's
Confessions. As it fell open he saw this passage: 'And men go about to
wonder at the heights of the mountains, and the mighty waves of the
sea, and the wide sweep of rivers, and the circuit of the ocean, and the
revolution of the stars, but themselves they consider not.'

FRANCESCO PETRARCH,
Familiar Letters[1]

I WENT TO A VERY traditional boys' boarding school. We rowed, we got fined if our top buttons were undone, and the most unmusical of us signed up to sing *Messiah* because girls from the local high school sang too. Prefects could carry silver-topped walking sticks, we were allowed to cycle after foxhounds at the weekends, and the honours boards picked out in gold the names of our ancestors who had carried the school's flame gloriously at Cambridge or the Somme. We learned Greek, beekeeping, fly-fishing, and, on the rifle range, how to shoot a charging Russian through the heart before he broke through the lines to disrupt the status quo. We had compulsory classes in what was called 'Community Wealth', which was an unblushing apologia for trickle-down economics. 'The best thing you can do for the poor,' we were told, 'is to become rich.' True religion was to have GDP as your household god. Real altruism was self-advancement.

One day a curious figure, dressed all in orange, with a Rasputin beard down to his chest, slank in scuffed desert boots into the school. All credit to the school for letting him in. He was fresh from the Poona ashram of the discredited yogi, Bhagwan Shree Rajneesh, and wore the yogi's photo on a string of beads around his neck. He'd been a boy at the school himself, had had a distinguished career at Oxford, and was a peerless sportsman. I suppose the school thought that anyone who'd made a century at Lord's couldn't be a bad egg.

His first move, after pledging to restore the flagging fortunes of the First XI, was to establish a brazen competitor to 'Community Wealth'. It was called 'Who Am I?', and it was an instant and embarrassing success.

Every Wednesday morning, for forty-five minutes, twenty bemused, hopeful, hormonal boys in tweed jackets, striped ties and grey serge trousers sat cross-legged and closed-eyed on cushions in a room above the business studies department, chanting the Gāyatrī mantra, going on astral journeys far from the sacred playing fields, fascinated by the sensation of their breath moving up and down their chests, called out of reverie by the teacher saying gently, '*Who* is feeling that? *Who* is seeing that? No point in feeling it; no point in seeing it, unless you know who's doing it, is there?'

He had an answer for everything, and it was usually, smilingly, 'No idea at all.'

It might not seem like it, but this is a book about living well. Living well, that is, as a human. Living well as a hedgehog might involve some of the same things as living well as a human, but there are some important differences too. Hedgehogs like eating slugs and, so far as we know, don't much appreciate string quartets.

To know how to live well as humans, we need to know what sort of creatures we are, and what sort of place we live in.

If we don't know *what* we are, and to what we might and should aspire, we might (even if we're content, or, more likely, *because* we're content) be letting our lives go off half-cock. We might be wringing less out of our threescore and ten than we could, and might be doing ourselves harm. We might be designed to run on unleaded, but be filling ourselves up with diesel.

If we don't know *where* we are, we might be responding inappropriately to the cosmos. We might be wandering round naked when it's freezing cold, and saying that our frostbite is unavoidable wear and tear. We'll be better off if we know it's cold, and put on a coat.

The study of humans is called anthropology. It is a strange subject, because it makes no attempt to say what humans are. Imagine a course on rodents that didn't begin by defining 'rodent'. And imagine a course on rodents in deserts that didn't say what was meant by 'desert'. Yet that's what we do with the study of ourselves.

Perhaps I'm being unkind to anthropologists. Perhaps there isn't a *failure* to define 'human', but a *refusal*. And perhaps there's a refusal because anyone who looks closely at humans is struck dumb with awe by what we are, and realizes that any effort to say what we are will be confounded by our complexity, our variety, our goodness, our depravity, our sheer category-busting inconsistency, our vertiginous giganticness, and our piffling littleness.

When we have tried seriously to say what we are, we have, for most of our history, reached for theological language. But theology tends to be better at posing questions and giving ethical guidance than at solving ontological conundrums.

'What is man, that thou art mindful of him?' asked the Psalmist. A very good question. The writer purports to answer it, but doesn't. God 'made [Man] a little lower than the heavenly beings,' he goes on, 'and crowned him with glory and honour ... '[2] It is no answer at all, for it says not *what* humans are, but *where* they are in the hierarchy of being.

It is useful, though. It tells us how to comport ourselves (with quasi-divine dignity) and how to behave towards other humans (you mustn't see a creature crowned with glory and honour as an economic unit, or send it to work in a brothel or a merchant bank). That is worthwhile advice, to be sure, but it's not *ontological* advice.

Similarly for the notion of the *Imago Dei* – the idea we are made in God's image. Belief that I'm basically God-like will, if I have a respectful view of God, inhibit my self-denigration, make me denounce sweat shops, and keep me from a career in the oil industry. But since God is infinitely mysterious, to say that I am like Him tells me little about my nature except that I too am mysterious. That, mind you, is highly practical knowledge. Learning what to do with it is urgently important. Since creativity is one of God's defining attributes, the *Imago Dei* perhaps means that we are foundationally creative creatures. That, again, is hardly an ontological answer, but it is worth having.

Apart from theologians, the other people who say that they can help define us are reductionists of many shades. Their approaches are various. Some say: 'We are higher primates who share much of our DNA with flatworms.' True, but unhelpful. To identify the branch of the evolutionary tree on which we perch just says where we are, not what we are.

If you get marks for answering the question the examiners set, the other type of reductionist answer scores rather better: 'Humans are merely machines – cocktails of chemicals.' This has the merit of clarity, but the demerit of being at odds with everything we know and feel about ourselves and the rest of the cosmos.

I'm going to suggest (I shudder at my hubris) what humans might be, at bottom. In doing so I'm going to talk about some human attributes. There is, or may be, a distinction between attributes and substance. This was recognized by Aristotle, and disinterred in the Middle Ages to defend the doctrine of transubstantiation. Were the Eucharistic bread and wine really the body and blood of Jesus? Yes. Then why did they taste like bread and wine rather than human flesh and blood? Because they retained the chemical attributes of bread and wine, while being in substance Jesus's body and blood.

This distinction sometimes matters profoundly – in medical

ethics, for example. We might be able to agree that practical reason is a defining attribute of humans, but does someone stop being human if dementia or traumatic brain injury robs her of that attribute? No: she is still wholly human. Is a toddler, adept with Lego but at sea with *Logos*, human? Yes.

We'll return to this distinction, but for the moment let's park the thought.

We can know *some* things about ourselves as a species. We can, at least, reconstruct our evolutionary origins and see where we've come from.[3]

We can see how we've behaved in the past and behave in the present, and by noting how we have judged and judge that behaviour, conclude something about our moral nature. We can, for instance, record our metaphysical predilections, the ways we structure our societies, our aesthetic concerns, our epistemic queasiness and our desire for certainty.

The process of looking at our past to understand our present is valuable. Just as our childhood years determine, to a large extent, what we are as adults (and particularly determine our subconscious, which is overwhelmingly the most significant part of us), so our childhood as a species determines what we are – at least in the deepest, most constitutional parts of us.

We've seen already, in our visit to the campfires of the first humans in Europe, that we've spent almost all of our time as a species as hunter–gatherers. What are you? You're Upper Palaeolithic, with a very thin patina of Neolithic, modern and postmodern crud which will wash off in less than a week if you go off on your own into a dripping wood.

Four aspects of this formative hunter–gatherer time go to the heart of what we are, and hence how we should behave: wandering; consciousness and storytelling; a sense that this place is not all that there is; and symbolizing and the use of metaphor. Each, as we'll see, has some connection with our intrinsic edginess.

WANDERING

We are built to walk and, since we are higher off the ground than most other mammals, we have a grand perspective of the land. In some senses we understand the paths along which the wildebeest migrate better than the wildebeest themselves.

This perspective may have contributed to our tendency to generalize and theorize, because that tendency is one aspect of big-picture thinking. It has helped to make us map-makers, and so hinted to us that maps may be as real or (latterly) more real than the land itself.

Walking keeps our feet literally and metaphorically on the ground. Hunter–gatherers often walk long distances in search of things to gather and things to kill (following the migration of the herds, for instance), and must also know, in relation to the whole territory over which they wander, exactly what will be blooming and breeding when and where.

To settle is to die. To stop moving is deadly. If blood stops flowing through our coronary arteries we have a heart attack. If a river becomes too sluggish it starts to smell. That's what happened to us when we stopped wandering. Constancy, wrote Freya Stark, 'far from being a virtue, seems often to be the besetting sin of the human race, daughter of laziness and self-sufficiency, sister of sleep, the cause of most wars and practically all persecutions.'[4]

There is no neat divide between wandering hunter–gatherers and sedentary farmers. Some hunter–gatherers are sedentary for some of the year, and some ancient hunter–gatherer communities built vast complexes requiring the sort of planning and cooperation long thought to be the prerogative of settled peoples. Göbekli Tepe, in eastern Turkey, is probably the best example. But because the divide is not neat does not mean it is not real. It eventually became very real indeed.

In the Neolithic – an era of settlement, when humans started to be concentrated for most of their lives in centres – lots of things started to die and smell. The egalitarianism which typifies hunter–gatherer communities began to wither. If plant-hunting women bring in three-quarters of the calories, it's harder for gazelle-hunting men to pretend they're the only important ones, but if the men are out in the grain fields, leaving the women grinding the grain at home, it's easier for men to pretend (and believe) that they're the linchpin. We see the birth of chauvinisms, hierarchies, power-plays and them-and-us-es. Of many kinds. We see the burgeoning hegemony of supply and demand, and consequential violence (Steven Pinker is quite wrong in his notorious reconstruction of the history of human violence[5]). We see the exhaustion of natural resources in the areas around settlements and a swaggering, colonial attitude towards the non-human world – a world previously seen as sacred. We see the birth of politics as we know it now – which is a kind of death. As communities grew larger, trust and shame could not police communities effectively, and law, with all its jackbooted henchmen, goose-stepped in to do the job instead.

If you think this is hysterical overstatement, perhaps you'll take God's word for it. He makes no secret of his preference for the wanderer. That's the root of the Cain and Abel story: the settled agriculturalist's offering is less acceptable to God than that of the Bedouin who wanders with his flocks over the hills of the Near East. The first farmer is the first murderer. Cain killed Abel in a *field*. Violence and envy enter the world with settlement.

As we saw when we met some pilgrims in Chapter 6, every serious religious tradition distrusts stasis. The universal commandment is to leave wherever it is you've become comfortable and become a wanderer again; become address-less to have a home worth living in; he who seeks to save his life (at least with the help of a mortgage company) will lose it. Religions are schemes for becoming what we should be; and one of the things we should be is a hunter–gatherer.

One of the things I notice about wandering is that wanderers, unlike people sitting at a desk, are constantly crossing frontiers. I'll return to that.

CONSCIOUSNESS AND STORYTELLING

We obsessively use personal pronouns: '*I* see', '*you* are', '*she* is'. Though we can't hear the voices of Upper Palaeolithic hunter–gatherers, we can infer those pronouns from certain types of behaviour (such as the way the human form is represented in Ice Age art, and from funerary practices). The earliest representations of the human face (such as the Venus of Brassempouy, an ivory figurine about 25,000 years old, found at Brassempouy, in France) are emphatically non-generic. They declare either 'I' or 'Thou'. And you don't put someone's treasured artefacts alongside their corpse unless there is a 'they' who wants and needs them.

Pronouns are seismic. That we use them at all tells us a huge amount about what we are; that we use them in the way we do tells us even more.

They say that I believe I am real, you are real and she is real, and so they recognize the possibility of real relationship.

They imply a belief that there is a world beyond my own head, and that I can know at least some things about that world. They scream, 'I am not alone, locked up in the echo-chamber of my skull'; and, taken with a lot of other evidence, they begin to hint that relationship itself is what the universe is all about: what, indeed, it all *is*.

They show a real faith in agency, and it's a hop and a skip from believing in agency to believing in *responsibility*: to feeling morally obliged; to making us moral creatures.

How might our peculiar type of consciousness have arisen? (We have seen something of this already. Remember those shamans, dancing until the blood poured from their nostrils?)

There is no overwhelming consensus about how our behaviourally modern consciousness ignited, but many support (as at least part of the explanation) the thesis of the South African anthropologist David Lewis-Williams, who suggested, based on his studies of the San people of southern Africa, that shamanic voyaging and/or the induction, without such voyaging, of altered states of consciousness, had something to do with it.[6]

The idea (we've met it already) is that by dancing, dehydration, or the ingestion of hallucinogens, humans went, or felt as if they had, to other realms – perhaps represented by the world across the cave wall.

From the vantage point of those other realms they looked back at the world they had left and saw it with new eyes. They saw themselves as distinct – generating the insight that there is an 'I', and that therefore there is a 'You', and that therefore there can and must be an 'I–Thou' relationship.

There have been almost no significant developments in consciousness studies since the Upper Palaeolithic – and certainly not since De Quincey was eating opium and Humphrey Davy inhaling nitrous oxide and William James trying to make sense of it all. Nobody has been able to suggest any reason whatever why natural selection should be remotely interested in promoting our type of consciousness. You don't need consciousness for anything relevant to our mere survival as a species. It's useful to have Theory of Mind, of course: to know what's in someone else's head. But you don't need consciousness for that. Consciousness is no use to Darwin. All it is good for is facilitating a sort of relationship invisible to natural selection.

Once you have pronouns, you necessarily have stories. We compulsively feel that the 'I' has to be situated in the cosmos, and we tell stories to situate it. Perhaps this is a consequence of the dizzy disorientation we feel if we don't know where we are or where we come from or where we are going – a disorientation which *is* the modern human condition, unmitigated by the comfort of religion or any other kind of metaphysical commitment.

A SENSE THAT THIS PLACE IS NOT ALL THAT THERE IS

The novelist Alan Garner said that 'everything humans do is necessarily votive ... to live as a human being is in itself a religious act.'[7] I agree. A belief that there are worlds other than the workaday one is a hallmark of behaviourally modern humans. If you see grave goods, you have behaviourally modern humans. If you don't (with a small caveat about Neanderthal burials), you haven't.

Shamanic touring in other planes probably contributed to our understanding of the sort of animals we are and the sort of place the universe is, for travellers get to know well the places they've left behind. Not all hunter–gatherers are shamans, but the average hunter–gatherer lives more of a shamanic life than I do. In their world everything is porous. Other things and other worlds bleed into humans and humans bleed back. A gossamer-thin veil separates this reality from other realities which may be more solid than this one. Early humans knew about ecology long before we laboriously described it. They lived it. They knew about quantum mechanics: they breathed and ate the equations.

Paul Pettitt, one of the greatest authorities on Upper Palaeolithic religion, argues that the standard Upper Palaeolithic belief was that our agency increases rather than decreases when we die.[8] It's of a piece with Platonism; of a more solid world of the Forms; and of the Christian teaching about the resurrection body of Jesus – a body far more solid than our vaporous bodies; a body which could pass effortlessly through walls – across edges.

SYMBOLIZING AND THE USE OF METAPHOR

Walk along the human evolution galleries of any decent museum. Start walking along the cases towards *us*, but beginning before *our* beginning. Look particularly at the artefacts. They are really boring until you get to the Upper Palaeolithic – to *us*. You'll know when *we've* arrived, because suddenly there's art – an explosion of *symbolism*: of solid metaphors; metaphors you make with your hands and pass down to your children.

Just think of the valency, complexity, possibility and agency this injected into the world. If, sitting in a cave with your flint knife, you can turn that piece of mammoth ivory into a lion-headed man which, as well as being both a lion and a man, is *still* a piece of ivory, is anything impossible?[9] With the edge of a knife you can make an infinite number of other edged things. Faces emerge from the ivory. Surely they were there all the time. If you scrape hard enough perhaps you'll find your dead father inside.

You can make anything into anything else. The universe is an intrinsically alchemical place and we're all alchemists.

You're a sort of god. Gods don't die, and they can make company for themselves. Everything – not just your father – seems to be immanent in everything else. So is death itself such a big deal? Life might be huddled inside the grave.

The mammoth ivory is a piece of solid metaphor. We are obsessively metaphorizing animals.

If anything can be anything, there is an infinite number of edges.

Those first modern humans were far more modern – far edgier, more inherently restless – than us. They walked constantly. If they were in Europe, they lived on the edge of the ice. They were often on the edge of hunger, viability and comfort, always on the edge of life, and always on the edge of new ways of conceiving the world. The earliest art we have is the edgiest art of all: it depicts shamanic voyaging and shapeshifting. We are the children of lion-headed and horned men. We went to the caves, pressed our hands against the cave wall – pressing out from ourselves towards the world on the other side to which the shamans went – and blew red ochre all around our hands so that when we were gone part of us (the aspirant, out-pressing part) would survive as part of the rock.

There's a fair degree of agreement about what we need in order to thrive. Decent cultures recognize that agreement in their norms and laws.

Martha Nussbaum, following Aristotle and just about everyone who has wondered systematically about human flourishing, has identified ten core needs.

Life is the most fundamental – the bedrock on which all the others are contingent. The others are *bodily health, bodily integrity, use of one's senses, one's thought and one's imagination, emotion* (including the ability to become attached to something other than oneself), *practical reason* (including knowing what is good and what is not, and cohabiting comfortably with one's conscience), *affiliation* (including the ability to be empathetic and sympathetic), a relationship with *other species* in the world of which humans are just a part, *play* (life isn't all about measurable productivity, and laughter has been hugely important in human evolution), and *control over the environment* (including political freedom and the ownership of adequate resources).[10]

Nussbaum casts her net wide. There is little she does not catch. While the list might seem rather trite, it provides, at the least, a set of pigeonholes in which to file our thoughts about the things we need to thrive.

Her list is easily transformed into a set of defining human attributes (for surely we thrive best when we act in accordance with our true natures). Use it when next, at a party, someone asks who you are. Declare: I am an embodied, sensual, thinking, imaginative, emotional, moral, social, playful, controlling creature, and I have a crucial relationship with the non-human world.

Nussbaum's list is often used in the academy. I've cited it hundreds of times, more or less unreflectively. But something odd set me thinking about it more seriously.

I was sitting in a snug church in Yorkshire. A property developer in a suit stretched by steak and kidney pie was reading, without embarrassment, the Sermon on the Mount. 'Blessed are those who mourn,' he boomed. 'Blessed are the poor in spirit.'[11]

I listened, bemused, as if hearing it for the first time. What on earth could it mean? Or perhaps it meant nothing at all on earth, but contained truths for another realm altogether. If it did, I thought that other realm was a desirable place.

The vicar, with an eye to the church roof fund, didn't preach on the day's text. Nor did he rudely mention the Magnificat's insistence that the mighty would be put down from their seats, the hungry filled with good things, and the rich sent empty away. But I was horrified.

The dispossessed, apparently, were the only ones who possessed anything at all. The moral centre of the cosmos was in the shanty towns outside the financial and cultural centres, in the hospices, and amongst the children scavenging on the smoking dumps outside Fortaleza.

What would Martha Nussbaum make of it all? Here were categories (the mourners, the poor, and so on) who at first blush didn't seem to be captured by her list. And not only were they said – with a shrillness that appalled and fascinated me – to be flourishing, but to be the only ones who were truly flourishing at all. I was back to that story about St Francis. Wasn't it the profoundest sort of perversion to worship suffering and poverty? What sort of life-denying deviant makes the funeral parlour the normative human experience?

I walked away from the thought, as most of us walk from all important things. But the thought pursued me. It raised its insolent head whenever I was feeling comfortable and satisfied: at a good dinner; at a reassuring diagnosis; by a cosy fireside; when all was balm with the family; when someone said something nice about one of my books; when the bank account swelled. I resented it very much. It was stealing my joys and denigrating things I thought were good. I wanted Martha Nussbaum back.

Had the thought been only Christian, it would have been fine. I could and would have dismissed it as a hysterical artefact of febrile first-century apocalypticism. But it wasn't. It was there, and strident, in the prophets of Israel, the mendicants of Hinduism and Buddhism, the scepticism of Lao Tzu, and in the sacrificial generosity enjoined by Islam. It was everywhere. And it was there too in my intuitions.

I didn't like it at all.

Then things took a different turn. For another project I started to re-read lots of old stories – stories I knew from childhood and had read to my own children. And I began to see that I had often not seen the wood for the trees.

The old stories weren't intended just to titillate or to make long winter evenings bearable. They were meant to say something fundamental about the sorts of creatures we are. They were meant to answer the Psalmist's question; to do the work, but more fundamentally, that Nussbaum had begun.

I knew about this aspiration to answer the Psalmist's question. I had assumed, without looking hard, that the old stories failed – really because the brief was such a big one. As I looked more closely, though, I wondered if the failure was as complete as I'd thought.

Many of the stories gave an account of origins and status. Many more described and applauded the qualities in Nussbaum's list. But they did so not by enumerating desirable or quintessential characteristics; not by piling up propositions, but by telling stories.

So here was my great insight: *stories tell stories*, and by doing so they do something very different from simply distilling the principles from stories and filing them in a cognitive cabinet. This insight, which is blindingly obvious to small children and all intelligent readers, had passed me by. That's what academia does to you.

Having seen this, there was no stopping me. Why could stories do what the distilled principles of stories could not? The stories, it seemed to me, were real in a way that principles were not – even if the stories concerned fanciful worlds and outlandish creatures. They had bodies and relationship at the core, and they moved through time. Stories, like us, have a beginning, a middle and an end (by now there was no limit to my excitement with the self-evident). They told us more about us than Nussbaum's list because we ourselves are stories, and it takes one to know one.

But *what* did they tell us about us? That needed lots of re-re-re-reading, for I am fogged with presumption. But slowly the fog cleared in places. What I saw was very, very strange indeed.

There in the stories, as I'd already noticed, were Nussbaum's characteristics, cheered on, expounded and advocated. We should all become heroes and avoid being anti-heroes. That was to be expected. I'd seen that all along. But where were those characteristics shown to their greatest effect?

On the actual and metaphorical edges of life. Knights went out from the comfort of courts and got lost in woods, and only there were they able to be properly knightly. When all was lost, and only when all was lost, were there great riches. It was all of a piece with those disturbing lines from the Sermon on the Mount.

But surely, I thought, this was just a literary device? Stories have to be interesting to hold listeners' attention, and extreme situations are more interesting than accounts of an OK day at work. But no, it wasn't that – or certainly wasn't *just* that. These old tales were saying something normative about what we are when all the pretension is stripped away, and so how we should act *at all times*.

They hinted strongly that, if we could only see it, we are *always* lost in a wood, or teetering on a precipice, or destitute, or the only one standing between a dragon and a fair maiden chained to a post, and we should behave accordingly.

My mind slipped to a moment years before in Varanasi, northern India, where an emaciated beggar in saffron robes held out his bowl to me, and when I'd walked past him without giving anything, screamed 'I am you'; to my discomfort when a friend asked me to consider giving a kidney to a stranger; and to yet another snug northern church where the vicar told the story of the man who poured all his money into building bigger barns, only to learn that 'this very night your life will be demanded of you.'[12] 'And God said to him: *You fool!*' shouted the vicar, in a strong Blackburn accent, and I wanted to crawl out into the dark. I could see all too clearly what I was, and didn't like it, and how I should thrive, and I wasn't up to it.

Nussbaum's list, then, was embedded in the old stories, but they said better than she did what we were, or at least identified better some of our attributes. One of the reasons for this was, as I've said, that the stories all involved relationship – with people, with places, with ideas, with malevolent lizards, with gods; with the world, that is, outside the heroes' skulls. Nussbaum had identified engagement with that world as central to human thriving and human definition. We simply couldn't consider humans (either as a species or as individuals) in isolation. We aren't brains in vats or even mind–body–spirit unities in vats. We bleed into the world and the world into us. And so to know what we are, we need to know what is being transfused.

That's a big ask. So big that few of us bother to ask – at least in a systematic way. The nature of reality: discuss. Did the old stories themselves have anything to say about it?

Indeed they did.

The very old stories and the very new stories said the same.

Here's a new story.

Albert Einstein and Niels Bohr had a famous argument. In principle, said Einstein, we could formulate a theory that would predict how everything in the cosmos behaved. Not so, replied Bohr: the behaviour of everything is affected by the observer.

We now know that Bohr was right. It means that the cosmos is at bottom a relational place. It is one colossal conversation in which everything, from sub-atomic particle to planet, is speaking to everything else, listening to everything else, and being influenced by everything else. In that sense, everything is one.

We needn't rest that conclusion on the language of mystics who, after years of meditation, talk about becoming one with the universe. We can derive it from the fact of quantum non-locality, demonstrated beyond doubt by clever people at CERN. If two bodies have been close to one another they will affect one another instantaneously, however far apart in time and space they are.[13] That's not because they send messages to one another, infinitely faster than the speed of light. It's because they are part of the single great organism that we call the universe, or, if you prefer, reality. Why? Because of course, at the moment before the Big Bang, everything *was* close – infinitely close – to everything else. Relationship is a far more fundamental property of reality than the forces binding atoms together. Forces are themselves just examples of relationship.

But while everything in that sense is one, everything in another sense is many – as Iain McGilchrist has powerfully demonstrated.[14] There is no contradiction here. Consider a magnet. It is an undivided whole: one magnet. But if one of its two poles is cut off, are you left with one pole? No: distinct north poles presuppose distinct south poles, and magnets demand both. Oneness depends on variegation.

We rightly insist that our experiences are our own. This declares our conviction that we are one in a universe of dazzling otherness. Our experience of beauty depends on variety. We might look up and say that the perfectly clear blue sky is beautiful, but if everything were an undifferentiated, seamless, contourless blue it wouldn't be beautiful. Beautiful blueness depends on un-blueness for its beauty. The cosmos isn't like a pint of homogenized milk. That's not what oneness means. Still less is the cosmos like a typical British high street, identical to every other high street. It's an incontinent generator of variety; it spews individuals.

Increase the number of individuals and you increase the total length of *edge* and the number of distinct viewpoints.

The cosmos is an edge-generator, and so a viewpoint-multiplier.

If the cosmos is a conversation – a festival of otherness in a vast marquee of unity – then everything, from electrons to elephants and beyond, is on the edge of itself, looking over.

There is an obvious way to increase edginess. A little geometry and a little biology may help to explain. Why do big animals do well and small animals do badly in cold conditions? It is because the surface area of a sphere is $4\pi r^2$. That *squaring* keeps shrews from the South Pole.

The greater the volume, the smaller the ratio of the surface area to volume, and so the smaller (relatively) the area over which vital heat can be lost. A musk ox has less surface relative to its volume than a shrew. A shrew is *edgier* than a musk ox. It impinges relatively more on the outside world. It is more vulnerable. Its surface *relates* more to everything that is outside. Its spleen is closer to everything that is not the shrew than the musk ox's spleen is to everything that is not the musk ox. A shrew, then, is a more relational entity than a musk ox. If the cosmos is made of relationship, a shrew is more cosmic.

Edges facilitate vulnerability and relationship. If you want more relationship in your life, get to know lots of small things rather than a few big ones. And be a small thing yourself. If relationality is what the cosmos is really about, edges are a proxy and a prerequisite for relationality. And the greater the length of the frontier, the more possibilities for incursion there will be.

Everything we know that really matters – that goes to what we think we really are, and to what we truly value in the world – we know by encounter: by experience. Would you feel safe in the hands of a doctor who had read everything there was to read about lymph nodes but had never palpated one? Would you listen respectfully to a lecture on the *Iliad* from someone who had never walked through the Lion Gate at Mycenae? Phenomenology trumps every equation. Data give way to intuition. What's between the lines is more important than what's on them.

Scientists can tell us things like our weight, or our pulse rate, or the species of flower we're admiring, or the origin of the rocks in the mountain we worship. But a complete accumulation of information about me, the flower or the mountain won't begin to account for the value I ascribe to my life, or the flower, or the mountain. It won't say what they *are.*

Recall Eustace, in *The Voyage of the Dawn Treader.* 'In our world,' he told Aslan, 'a star is a huge ball of flaming gas.' Not so, says Aslan. 'Even in your world, my son, that is not what a star is, but only what it is made of.'[15]

McGilchrist suggests that the palpable poverty of the Anglo-American tradition of analytic philosophy (a poverty which makes it aridly irrelevant to most of life) is a consequence of the absence in English of different words for 'to know.' The French have *savoir* and *connaître*; the Germans *wissen* and *kennen.*[16] But we Anglophones are forced to pretend that they are the same thing, which they obviously are not. Both are vital. Both should inform one another. But if I had to sit beside one of the verbs at dinner, let alone be married to one, give me *connaître* or *kennen* every time.

Connaître and *kennen* presuppose leaky borders.

McGilchrist tries very directly to answer the Psalmist's question and to say what sort of place we inhabit in his immense book *The Matter with Things*, in which he breathes neurobiological life into Heraclitus and Hegel.[17] What appear to be *things* – entities – are (he argues) in fact *processes*. That includes us. Consider a wave: in a sense it is distinct from the sea.[18] It has its own shape. That shape is different from the shape of other waves. I can describe its discrete behaviour. Yet (without in any way denying its individual wave-ness) it is part of the sea. It is sea water. It is part of the history and the future of the sea; part of the way the sea is unfolding.

What we think of as 'things' are like waves: episodes of unfolding. This unfolding is not like the opening of a fan, which will expose the picture that has been on it all the time. There is plenty of room for contingency, agency and creativity.

McGilchrist contends that consciousness is ubiquitous, and not just a property of the aggregation of neurons. He is joined by a growing army of philosophers and physicists.

Here is the problem for the materialist: how can consciousness result from unconscious matter? The simplest response – which begs fewest questions – is to say that matter is not unconscious at all. Philosophers such as Alfred North Whitehead, Thomas Nagel and Galen Strawson have defended this position. Many physicists see this conclusion as an inevitable inference from Bohr's demonstration of the influence of observers on all observed things and phenomena. If consciousness is everywhere, chatting with and sparking off other consciousnesses, poor old determinism is indeed dead – as our instincts always knew it was.

McGilchrist draws deeply on the French philosopher Henri

Bergson, who saw time as like a snowball rolling down a hill, accumulating snow as it goes but taking the past with it.[19] The past is immanent in the advancing ball, but there is nothing deterministic about the roll. I prefer Jenny Odell's take on Bergson: time is like lava flow over relatively level ground; the leading edge of the flow is alive and dynamic.[20]

There's no room in these conceptions of time for the debilitating notion of time as akin to a series of steadily chugging railway carriages, each carriage representing a moment; or as a line, with the past at one end, the future at the other, and the present at the middle. McGilchrist cites the physicist George Ellis, who (consonant with snowballs and lava flows) envisages an evolving universe of space–time whose volume grows as it rolls. The present is at the surface. Since we're temporal animals, so are we. We're at the edge of the snowball, at the edge of the lava flow, looking out into the void.[21]

This accords with our sense of our selves: with the precariousness we feel when we bother to consider our predicament seriously; with the scariness of looking up and out (the direction of gaze forced on us by our upright posture and our curious cognitive constitution and by the constitution of reality); with our feeling about the salience of the moment and our feeling that the past is always present; with our swashbuckling adventurousness and our timorous conservatism; and with every single one of the really good old stories about forests, mountain trolls, treasure troves and journeys to outlandish lands.

It accords, too, with what we've seen from our hunter–gatherer past. It accords with our knowledge of ourselves as wanderers; with our consciousness and the individuation inherent in 'I' and 'Thou'; with our sense that there are other worlds out there at the edges of this one, and that there is commerce between the worlds; with the suggestion that everything is composed of seams between worlds; and with the infinite individuation that comes from symbolism and metaphor.

*

We see life best from the edge of life and comfort and conscious-ness. The best literature is from the dispossessed. As we've seen, Salvador Dalí actively sought the zone at the edge of sleep and wakefulness where he found the visions which fill his paintings. We see the things that matter only at the edge of vision, and between, never on, the lines of poems and novels. In the Jewish tradition, the nation of Israel was redeemed and shaped by going over the Egyptian national boundary and surviving in the desert, well over the edges of probability. In the Christian tradition redemption is effected by the death of an edge-man – edgy from his conception onwards – on a hill outside the metropolis. The Prophet Muham-mad had to flee Mecca and the established orthodoxies to establish the new religion. In the religions of the East, dispossession of the self *is* enlightenment.

And so I think I begin to see something of what those strange verses in the Sermon on the Mount might mean.

Reepicheep, the gallant mouse in C. S. Lewis's *Voyage of the Dawn Treader*, is excited when the sea water turns sweet, for it means the edge of the world is near. He refuses to go back with the others, but paddles furiously in his little coracle towards the edge. We last see him at the crest of a wave – a wave which, like time and reality itself, pushes on towards new possibility. It's as if he's become part of the wave. Bergson would approve.

Graham Greene, one of the supreme chroniclers of the human condition, who made a pretty fair stab himself at the Psalmist's ques-tion, said he'd choose as the epigraph for his novels an observation in Browning's 'Bishop Blougram's Apology': 'Our interest's on the dangerous edge of things. The honest thief, the tender murderer, the superstitious atheist.'[22] Greene was a great novelist because he knew that normal thieves are honest, normal murderers tender, and all atheists superstitious.

There's nothing we can do about our edginess. It's just the way it is. We live in a cosmos characterized by edges, constantly broadcast-ing and receiving over the boundaries of ourselves, for relationality *is* the fabric of reality. Edges are of the essence. The present we inhabit is a rolling edge with, within it, all the rolling edges of the past.

The sixth- and seventh-century theologian Maximus the Confessor, who went over the big edge in Georgia in 662 CE, not long after his tongue had been pulled out and his right hand hacked off for his supposed Christological heresies, gave the most complete answer I know to the Psalmist's question. Each human was, he held, a microcosm of the entire universe. There's a cosmos in each one of us. This explains our colossal moral significance. It chimes with the Talmudic dictum, made famous in *Schindler's List*, that to save one life is to save the whole world. If, as I've argued, the cosmos is made of edges, we embody all the edges there are.[23]

Each of us is a cosmos. Each of us looks out at a multitude of cosmoses. What do we see? Do edges have anything to do with our view?

11

Viewpoints

The experience of exile, of being a diaspora, was in many ways a liberating experience [for Russian Orthodox thinkers who found themselves in Paris after their expulsion from Russia in 1922]; *the engagement of these thinkers with the Western thinkers who welcomed them led to a rediscovery of elements of Orthodox theology that had been forgotten or overlaid in the immediate past . . . '*

ANDREW LOUTH,
Modern Orthodox Thinkers[1]

WHEN SHE WAS ABOUT five, I took my daughter Rachel to London. Well, I took her to some places *in* London.

One of those places was Parliament Square. She looked around, pointed to Big Ben, which she'd seen on a biscuit tin, and said, 'Is that London?'

'No,' I said. 'That's Big Ben. It's part of London. It's in London. So are we.'

'But where's London? I want to see London. You promised me I could see London.'

'You can't actually see all of London at once. You'd have to go up in a rocket and look down at it from space.'

'So you weren't telling the truth. You said you always told the truth. You broke your promise.'

'No, I didn't. That is – this is – all part of London. Just not all of it.'

'Then take me to the middle.'

'This is the middle.'

'Well, if this is the middle, why can't I see all of London?'

'I suppose that the middle is the place from which you see least of London.'

'That's silly. You don't see more of something if you go further away from the middle.'

Usually my children are right, but actually, you do, Rachel. If I'd wanted to show her more of London than here, right in the centre, I could have taken her up Highgate Hill and on to Hampstead Heath, or up in an aeroplane.

When she's older I hope she'll be troubled by the fact that we're governed from the place which has the worst possible view of the centre, and no view at all of anywhere else.

On a drizzling November evening, on a roundabout on the outskirts of Hartlepool, a Ford Fiesta, driven by Dave, ploughed into the passenger-side front door of a Vauxhall Zafira, driven by Jane. Jane said she had mild whiplash. The respective insurers couldn't agree, and the case, as a tiny proportion of such cases do, found its way to a court, where a judge had to decide who was to blame. There were no witnesses other than Dave and Jane. They blamed each other. They couldn't both be right. The site and nature of the damage to the vehicles, and to Jane's cervical spine, didn't help at all.

It was a common but difficult judicial exercise. It would have been much easier if there had been other witnesses. The more the better. If the judge could have had three hundred and sixty witnesses, one from each degree around the site of the collision, and if each of them had been watching the scene for five minutes before and five minutes afterwards, the exercise would have been trivial, if very boring.

Dave and Jane, in the cars at the centre of the drama, were the worst-placed of all these potential witnesses to comment on how the accident happened. They were too close to it. They weren't geometrically – let alone psychologically – capable of objectivity.

Multiply your viewpoints and place them outside the centre, and you're more likely to understand. That's the gist of post-structuralism. For we can only understand in our capacity as humans. Epistemology is constrained by anthropology. Humans are inherently indeterminate animals who are, or should be, constantly on the move. Decent epistemology has to take this into account.

Reeve and Kerridge, writing about the poetry of the British poet J. H. Prynne, say this:

> Post-structuralist theories of the decentred, destabilized self, moving along a series of subject positions constructed by discourses, rather than occupying a single, external viewpoint, and always confronted by excess and lack, seem to describe the reader of Prynne rather well. The apparent impossibility of achieving a complete reading of a Prynne poem, a reading which exhausts the poem's otherness, suggests that the poetry is postmodern in its indeterminacy, its avoidance of totality and closure . . . [2]

As for Prynne, so for everything. As for metaphorical travel, so for physical travel.

The point of travelling is to understand. You have to cross frontiers to travel properly. Often, though, we travel a thousand miles or more and don't cross any frontiers at all. Sometimes that's the world's fault. You might honestly want a new vantage point, but find yourself in a place identical to the one you've just left: a hotel owned by the same shareholders who own your home city; a road befouled by burger joints spewing the same food you eat back home. But that's not the worst type of non-travelling, for you might always encounter the wild, untameable, boundary-transgressing bacteria on the hands of the chef in the chain hotel, which might colonize your dinner and leave you hunched over the toilet bowl having a genuinely tropical experience, unmediated by an imported prescription.

No, the worst sort of non-travelling is where you take the whole of your world with you, buried deep and invulnerable inside – so deep it can't be dislodged by the most florid case of prickly heat in your groin. It's best to travel alone, so that your friends won't insulate you from the place you're in; won't keep you in England; won't stop you crossing edges and climbing on to new edges, looking back, seeing what you'd missed before.[3]

When we travel properly, most of the time we're looking back. We don't think of it like that at the time. We think we're looking around at temples or monkeys or baobab trees, but really we're climbing the temple or the tree for a better look at England, or seeing Scotland through the eyes of a rhesus macaque. We are locating ourselves more accurately by increasing the number of cross-bearings. We're hoping that with the distance and the new angle we'll be able to say – or say with slightly more confidence: 'Yes, that's *where* I am and *what* I am.' The great travellers amongst my friends don't have the faraway look in their eyes that sailors have in *Boys' Own* stories. They are secure where they are, because they know better where that is.

Assuming you're actually in a foreign place (rather than immersed in Mancunian gossip in a bit of Manchester that just happens to be in Bangkok), a good test to determine whether you're travelling usefully – whether you are actually on a new edge with a new view – is, unsurprisingly, to describe what you see when you look back. If it's the real thing, what you'll see is not what you miss about being back home (the comforts, the people, the reassuring routines, the hedges, the drizzle, the roast beef and Yorkshire pudding), but unlooked-for things; things that weren't in your line of sight before, or which were, but you didn't notice.

Shivering in a wind-blown tent on Svalbard, looking nervously for polar bears, I saw that a relationship was corrosive for us both, and that I had to cut and run.

Panting up a river valley towards the Tibetan plateau, I noticed that the stone by the bus stop at the top of our road in Sheffield could have been Neolithic.

Paddling along a brown river in Guyana, I knew I'd neglected a corner of the wood next to us in Oxford, and that there might be badgers there.

Wondering in the bush of northern Queensland if my bowels would ever return to normal, I observed that I had forgotten to forgive someone for a slight, and that the failure was doing me harm.

Pulling ticks out of my armpit in Ethiopia, I remembered I hadn't read *The Last Chronicle of Barset*.

Whisked through the frozen forests of northern Minnesota by a team of huskies, I realized why the swifts had chosen that hole rather than the other one to enter the nest site in the eaves over my study.

Sitting in a cave in Sinai, listening to belching camels, I knew that there wasn't much to me at all. I thought that when I got to Alexandria the busyness of the city would put me back together. It didn't, and that was where my own recent history started.

Sitting before dawn in a jungle ashram in Tamil Nadu. I realized that the bathroom tiles needed sealing.

It is always so. Not because we are self-obsessed or parochial, but because the world is one place, and Tamil Nadu is part of my bathroom and Minnesota part of my study, and because all knowledge is subjective and relational, and because I only see myself and anything else once I get out of myself. *Ec-stasy* again. That's why travelling is ecstatic.

All this is strikingly evident in diaspora art of all types. We have seen already how the provinces are overrepresented in the arts, and how the provincials remain provincial (retaining their edge-perspectives) even when they move to the geographic centres. Anyone who doubts the value of perspective, or doubts that perspective gives primarily a view of the self and of home (thereby rooting the artist and giving her a voice of her own), need only look at, for instance, the diaspora artists of the Indian subcontinent. India has the largest of all diasporas – 17.5 million in 2019, followed by Mexico (11.8 million) and China (10.7 million). The Indian names are glittering: Anita Desai, Bharati Mukherjee, Shauna Singh Baldwin, Amitav

Ghosh, Anjana Appachana, Anita Nair, Jhumpa Lahiri, Chitra Banerjee Divakaruni, V. S. Naipaul, and many more.

Naipaul can speak for them all. Though he lived most of his life outside India, he was always imaginatively there. He is the great chronicler of dispossession: a methodical geographer of the edges. He has been admitted to the centre because he has walked on so many edges, looking inwards, that he knows the centre (including the centre that is his Self) and its language better than anyone. He tricked the centre, by his knowledge, into thinking he was one of its own. This is a common history of edge-people. They don't inhabit the centre, but may convince others that they have a salubrious address there.

We need to leave in order to find. It is one of the foundational motifs of the cosmos. 'Whoever loses his life will find it,' taught Jesus. It was not just an ascetic or ethical dictum, but an epistemological proposition. 'Not until he left Venice, his known universe, did life begin,' wrote James Cahill of the eighteenth-century Italian painter Giambattista Tiepolo.[4] For Tiepolo the Venetian sky was 'not a screen but a threshold'. It was in South Africa that Gandhi realized what India was and might be, and that he himself was inalienably Indian. '.. [W]hat should they know of England who only England know?' asked Kipling – who was born in India.[5] He had a point.

The exiled Jews wept beside the rivers of Babylon, but they did not only weep: the core of the Hebrew scriptures was composed there. It was there, nine hundred miles from Jerusalem, that the myth (whether historically true or not) of Israelite nation-making was systematized.[6] In Babylon the First Temple, just destroyed, became bigger, holier and more golden, presiding over a Golden Age. Its destruction demanded strenuous theological justification, and Jewish theology was invented. It was there too that the mythos of Jerusalem itself was re-organized and Jewish identity and ethics reconceived in relation to the physical land from which the Jews were exiled. Many of the 613 *mitzvot* – religious obligations – can be performed only in the land of Israel. Distance and new perspectives are very creative.

Sometimes they can create the very notion of a centre, as they did for Jerusalem.

This special status of Jerusalem, inherited by Christianity from Judaism because Jesus the Jew died there, was to have curious later effects. On medieval maps, Jerusalem is the centre of the world. It meant that everyone who did not live in Jerusalem was a provincial – an edge-person – looking at the story of salvation, which was the story of everything, from the margins. Some tried to replicate Jerusalem wherever they lived. Throughout Europe there are round Crusader churches recalling the Church of the Holy Sepulchre in Jerusalem, and there was a brisk trade in relics from the centre, but the churches and the splinters of the True Cross, if anything, emphasized the fact of exile; emphasized the marginality of non-Jerusalemites. That sublime reliquary, Sainte-Chapelle, built in the thirteenth century to house the Crown of Thorns, is the most French building in the world apart from the Eiffel Tower. It is emphatically not in the Holy Land. I suspect that this marginality accounts in part for the vigorous bloom of medieval theology. Exiles are energetic and innovative. Jerusalem itself, even for Jews, has not been particularly productive. It inspires longing and praxis, but not the speculation that makes for great literature or art. It would be almost sacrilegious if it did, for Jerusalem is (in some ways) complete:[7] it's the centre; it's the point of things. The Jerusalem Talmud, a treasury of rabbinic disputation, is much shorter and less influential than the Babylonian Talmud.

We have visited some exiles who gaze from their edges with peculiarly clear vision at their homes, and so at themselves. It's worth seeing ourselves properly. To understand just how worthwhile, remember how vast Maximus the Confessor thought we are. He held that every human was both a macrocosm and a microcosm of the universe. It follows that every true description of oneself or any other is a description of the whole universe.[8] If the sullen, truculent youth at the café contains the cosmos, you'll be less likely to be as rude to him as he might otherwise deserve.

Beware, though. The geography of viewpoints from which we

can see things properly is rather strange. The truth is like happiness: it cannot be approached directly, but only from the side. We will see why shortly. It would be hard for me to write directly about Oxford – both because I am too close to it to see it properly, and because it is hard for someone who lives there to approach from the side. I could write about Oxford by writing a book about Namibia. That would give me a side view. I also feel reliably kinder and more forgiving about England when I'm abroad, and kindness is an infallible marker of better vision – whether of a country or a person.

I'm told, though, that there is no market for travel books, which is a shame. And apparently I shouldn't write about Namibia (or anywhere other than England), because that would be a kind of colonialist presumption. This is pernicious and chauvinistic. It says to the literary Hungarian visiting Oxford: this is *my* country, and only the English understand it. Don't you dare presume to tell me anything about it.

The best writing on Oxford is by the Chinese traveller Chiang Lee, who visited in the 1940s and wrote an incandescent book, *The Silent Traveller in Oxford*.[9] He saw the place far more acutely than I can. It's not surprising. I don't think his book should be pulped. The best evocation of Shropshire is *A Shropshire Lad*. Housman was born in Worcestershire, wrote many of the poems in *A Shropshire Lad* before he had visited Shropshire, and lived for most of his life in London or Cambridge – though he returned to Shropshire as a corpse, and is buried in Ludlow.

If we're ruled by the taboo prohibiting me from writing about anywhere other than me and mine, we can forget about nature writing, of course. What can a human say about the non-human world?

Exile, needless to say, doesn't *necessarily* produce better sight. It can warp as well as sharpen, producing saccharine sentiment; making England's grey and toxic land green and pleasant; sanctifying the unclean; defending the indefensible. But that doesn't alter the basic principle. You get a *different* view from edges, but not necessarily a more accurate one.

We can stay at home and still look over edges. They're everywhere. (Remember the village naturalist, tracking her snails?) Metaphorical edges are good. Physical edges are better.

Have you ever wondered why people retire to the seaside, leaving behind the places where they raised their families and earned their wages, and give their children's inheritance to a nursing home so that they can die looking through the double glazing over a cliff to the sea?

I used to think it was because the big open space of the sea reminds us of the great plains of East Africa, where we grew up as a species, and that the atavistic memory reassures us – comfortingly, when we are facing death – that we are at home. Well, perhaps. But I now think it's more about edges. As the mammoth ivory showed us in the last chapter, we are incurably metaphorical people, however literal we've been in our workaday lives. *Metaphorical* is a far more fundamental description of *Homo sapiens* than 'hairless, big-brained, bipedal ape'. And we get more metaphorical as we get more metaphysical, and we get more metaphysical the nearer we approach that state where metaphysics overtakes the physical: death. We are easily habituated animals. We get used to things quickly. And so (I think we intuit, when we're applying for our room in the seaside retirement home), if we choose to live in one big centrally heated metaphor for death, complete with tea dances and whist drives, real death won't be so scary when it comes. It'll be, if not a friend, then at least an acquaintance.

Perhaps there is something else. Madeleine Bunting, in her book about England's seaside, writes: 'I became intrigued by the idea of what gets exposed at the edges, what unravels and frays.'[10] Life itself unravels and frays in England's seaside gerontocracies.

Perhaps part of the appeal is to see how other people die, so that we can do it better ourselves when our time comes. Perhaps there's a comforting solidarity of the doomed (which we should all feel all the time with everyone, but don't). Perhaps there is something about seeing human life in the raw before it's too late. Perhaps the typical coastal mausoleum is really an anthropological museum where we can see what sort of animals we really are, and where we really live. If we live at the fringe of a titanic metaphor which has two tides a day and a pier sticking into it, we're forced to confront ourselves. It makes us acknowledge that there's either nothing or a tank of riddles beyond us. Every wave brings a fresh dose of *timor mortis*. We can get habituated to it. Habituation is the point of an apartment behind the Edwardian facades by the esplanade, with a stair lift.

It's not just the moribund and depraved who go to the seaside. There are far more painters and wannabe novelists per square foot at the coast than inland. Why do families flock to the beaches and sit in makeshift shelters in howling sandstorms? It's certainly not because it's enjoyable. It's not. It's not because they like swimming. They don't swim. It's not because they can escape convention and have sex with strangers: they're there with the kids and everyone's tucked up in bed with their phones by half past nine. The only plausible answer is that deep down we feel the need to confront these edges and learn from them, however wretched and poor it makes us.

Ask people what they've got from their seaside holiday, and they all say something along the lines of: 'We're rejuvenated.' They're new. They've put the past behind them and moved on. They've remade themselves. There's something about staring at the pier through the pouring rain that does that. There is something, too, about the evanescence of maritime light and the unpredictability of the sea (which might soothe one moment and eat you the next) tempered by the clockwork regularity of the tide. These all remind us of the vagaries and vicissitudes of life itself, and of our bodies: the clockwork heart, but the constant possibility of an electric storm in the heart muscle; our moods shifting like Monet's sun on the Étretat surf he so manically painted.

In looking over the fringe of the land to the sea, we see human stories encoded, dignified and made poetical. We are flattered and frightened. Reflected by the sea, our lives are stranger and weightier, and connected with other lives, including the consciousnesses seething below the waterline. We mooch along the tideline, looking to see what the sea has disgorged for us, and thinking every bit of driftwood freighted with story. The sea gives, the sea takes away, and the taking-away of all things including our last breath becomes more bearable.

The dog likes it too, though the wolf he really is comes from an Arctic forest where the only catchable fish are migrating salmon. The sound of the surf that we find so restful is really the noise of things being smashed, eaten and reconfigured. We have the pulverized bodies of Silurian trilobites between our toes and in our sandwiches, and somehow that is the way things are meant to be. It tells us that we'll be recycled, and that this is a kind of resurrection. We see new things as we look out to sea and walk along the seam of the sea and the land, and we're reconstituted, as the trilobites have been. Time doesn't run here as it does at home. At home, Greenwich rules. It pips out the hour from the centre, but here you'll never hear it over the scream of the wind. The wind and the time here are true in ways that the pips are not, and there is some comfort in that.

We see the same thing in many contexts. There are no emotions in the anatomically peerless statuary of Greek art until Alexander broke up the centrist hegemony of Athens and empowered the outstations at the edge of empire. In the Hellenistic period, it's as if someone has breathed life into the marble: smiles and frowns erupt from the stone. Those provincial sculptors weren't looking back at Athens as home. It never was. With the enlarged vision of reality given by Alexander's conquests, they looked over the edges of conventions at themselves and their friends, and out across the frontiers of newly conquered territory, and were freed by the sight to say how humans really were – as varied as the Étretat light, and as moody and unpredictable as the ocean. They were rejuvenated by the look, just as the sandcastle architects of Worthing are, and their statues were juvenated.

Greek science and philosophy as a whole may be another

example. Nigel McGilchrist (Iain's brother) contends so in a very important book on Pythagoras.[11] He seeks to identify the reasons for the profound differences between Greek thought and the thought of the cultures near the Greek world – particularly Mesopotamia and Egypt. He suggests many, but they boil down to the fact that the Greek world was unpredictable in a way that the other cultures were not.

Look at a map of Greece. It is mostly edge. It has the third-longest coastline in Europe (after Norway and Russia). The ratio of its area to its coastline is 9.65.[12] For France, the corresponding ratio is 159.62.[13] Modern Greece is, and ancient Greece was, dominated by the sea. It is said that there is nowhere in Greece from which, if you climb a nearby hill, you can't see the sea. That's not far from the truth. Greece is mostly edges, and beyond, the sea. The Aegean and the Ionian seas are notoriously capricious and ravenous. They swallow young men. The weather can be sun-lounger blithe one moment and murderous the next. The ruling Olympians are as fickle now as they ever were. Every headland in Greece has a whitewashed chapel to deflect Poseidon's trident from the fishing boats.

Ancient Egypt and Mesopotamia were very different. They were not defined by the savage contingencies of the sea. They were riverlands, and in antiquity, though there were blips, the inundations came and went regularly enough, making the land fertile and feeding the people. Survival and prosperity were assured as long as the monarchies were secure, which for most of the time they were. It was partly the job of the priesthoods to be guardians of royal security, for the monarchs were divine. If god's regent was visibly on the throne in Thebes or Ur, why worry?

Egypt and Mesopotamia had another, related, source of security: sacred texts. These described the source of order in this life and the next. They answered metaphysical questions.

The Greeks had no authoritative sacred books: no Books of the Dead; no Pyramid texts. If a Greek wanted to know what was going to happen to him when he died, he couldn't just look it up or ask a priest. There were inchoate, confused, contradictory clues in the myths, but they were a far cry from the detailed schemes enjoyed by their neighbours. The nearest thing the Greeks had to a sacred text was Homer, and he wasn't near at all. He is as full of uncertainty as the Aegean. We are never sure how the *Iliad* or the *Odyssey* are going to end. We are never convinced, until the last minute, that Odysseus will get home to Ithaca, or, if he does, that Penelope's bed will still be his. The mistiness of the Shades in Hades is code for metaphysical haziness about the fate of the dead.

Nothing ontological was handed to the Greeks on a plate. They had to work everything out from first principles. The work had begun by the time Pythagoras was born, and made great progress with the pre-Socratics. Socrates, Plato and Aristotle had rather different inflections both from the pre-Socratics and from one another, but the nature of the project remained essentially the same: what are we and where are we in the light of the fact that we're poised here on the very edge of the world, looking out at the Aegean, which is licking its lips after eating our sons?

Viewpoints are of course not merely geographical, but since we have bodies, and have to be in particular places, geography matters profoundly. When being on a geographical edge is combined with other types of edginess, though, there is often extraordinary synergy. The viewpoints seem to be multiplied exponentially rather than merely added. Then we sometimes see very great artists indeed. Depression, if it does not kill or disable, may be powerfully catalytic.[14] So may altered states of consciousness of various kinds. And homosexuality, which has often banished men and women to the edges of society under the threat of violence, provides many dramatic examples.

Leonardo da Vinci is perhaps one of those examples. Born out of wedlock, outside the little *comune* of Vinci, itself outside the metropolis of Florence, he moved restlessly between disciplines – including painting, sculpture, anatomy, civil and military engineering, and natural history – and places. A legend says that he took himself out of human society altogether for a while, living in a cave somewhere near Florence. I like to think it was there, amongst the rocks of his Tuscan hillside, that he had the vision which became the *Virgin of the Rocks*, and for which he invented a wholly new mode of perception, fitter for the subject. He had invented or discovered, he wrote, 'another kind of perspective, which I call "aerial . . ."'.[15] It gave depth to the air itself. He knew that new perspectives sometimes meant breaching taboos. He choked down his repugnance and crept out at night to dissect human corpses.

I have often wondered whether, without Ioannis Argyropoulos, who may have been Leonardo's tutor, and who certainly influenced him, we would have had the *Mona Lisa*. Argyropoulos is the person I'd most want to have dinner with. He studied theology and philosophy in Constantinople, and when it fell to the Ottomans in 1453, fled first to the Peloponnese and then to Italy. To which edges, from his own tempestuous past, did he introduce Leonardo? I have inherited many of the insecurities of my parents and my most inspiring teachers. They are the greatest bequests; my most valuable and most resented possessions.

Leonardo's sexuality is much debated, but the Alexandrian Greek poet Constantine Cavafy was certainly homosexual, preferring grimy, unwashed men, as far over the edge of decorum as he could get. He lived over a bordello in Alexandria, in a house that never had electricity, perhaps because he agreed with me that shadows are more interesting. He was a Greek who didn't live in Greece, but didn't live in the Arab world either: he is said never to have entered an Arab house. He was a Greek Orthodox man who scandalized the priests and scoured the back streets for norms to violate in his bed and his verse. He sat in the dark, or in candlelight, listening for the tread of the latest young mechanic he'd paid for sex, looking towards Greece over the sea, and seeing it better than anyone has ever done, because he saw it from so many more places.

And supremely for me, there is someone else.

I had been on Millbank, eating a pie and watching gulls

squabbling over the corpse of a rat. It was very interesting, and it took me a while to notice it was raining. When I did, I had an hour to kill, and took refuge in Tate Britain. It's not my favourite gallery. It's too preachy for me; it seems to deny the paintings' own ability to speak. A gallery should have the confidence to do without interpretative captions. But today one painting didn't just speak, but screamed, and all the mumbling in my head and on the explanatory panels fell silent.

It was called *Repose on the Flight into Egypt*. The Holy Family rests in the shadow of a vast recumbent statue (or perhaps it is not a statue at all). Mary, Joseph and the child are attended not by the shepherds and the farm animals of the nativity, but by a mélange of mythic figures: a black-winged sphinx with a diadem blazing in its forehead, centaurs, a faun, and a naked young man, wholly human but with a rear-end stance like a pony. They do obeisance to the family, but equivocally. They are more interested than worshipful.

I'd never heard of the artist. He was Glyn Philpot – a comically suburban name. When I got home, I rang round my artistic friends. They generally hadn't heard of him either, though one said: 'Yes. Queer. Portraits of black lovers. Not bad.'

Not bad! He could do anything. His early portraits, I was to discover, were distinctly Old Masterish, and he acknowledged his debt to Titian and Velázquez in particular. But there is a maudlin furtiveness even in his society portraits. Dandies are given the dignity of introspection. Grand ladies are rotting under their finery, and know it. Their eyes never meet ours, because that would be overkill: there's quite enough disclosure in the angle of the head or the line of the jaw. If one of the subjects turned to us, we would go mad. We do not have the software to process what we would see there. These pictures speak of the immensity of humans, the weight of even the tiniest gesture, and the poise needed to carry on breathing, let alone to sidle through the park or bury a child. Philpot invites us to follow the gaze of his subjects, which, though they look inwards, look outward too, for we are Januses. His recurring questions, which the sitters pose to themselves, are: where and what is the Self?

Could Philpot answer those questions? Did he think that anyone could answer them, for themselves or anyone else? If there were answers, they were as elusive as the light on Monet's beaches. He chased them far and fast, from Lambeth School of Art, to Paris, to the Roman Catholic Church (he converted as a young man), to a studio in Chelsea, to Spain, Venice, the United States, North Africa and Berlin. But the quest started in Herne, Kent, where, though born in London in 1884, he lived until the age of seven. 'Herne' comes from the Old English *hyrne*, which means 'corner', a place where edges come together, for this angular part of England juts out into grey sea.

I went there to try to find him.

Jeeves took his holidays in Herne Bay. I doubt he'd go there today, though he'd approve of the bungalows bristling with St George flags. People mostly drive well-waxed cars, but there are a few dog-walkers with spray-on trousers with no pockets for plastic bags because dog shit is someone else's problem. There are phalanxes of gnomes, and a whirr of strimmers. Whole ranks of shops are given over to funeral directors. Embalming sheds jostle potting sheds.

Suddenly there is the sea. We're on the eastern edge of England, but the sun won't rise for long for the denizens of the nursing homes here. Everything, from sewage to hope, drains down into the sea. Just off the coast, a wind farm carves up the air and the geese that fly in from the Baltic. On the high street it's chips with everything, and clouds of candyfloss.

It was tidier in Philpot's day, and more desperate. There were more poisonings and beatings behind the lion's-head door knockers than there are now. But it was a less articulated kind of desperation than today's; a desperation partly sublimated into the 1662 Prayer Book service at 11 a.m. on a Sunday, where they loved the story of Joshua's genocidal campaigns.

It's either too hot or too cold here, and sometimes both at once, for the wind from Denmark can freeze your neck as the sun from

Sittingbourne burns your nose. Two trains an hour go to London, but they might as well not.

There's a pier edging out to sea for significant things like getting engaged and scattering ashes and worrying about scan results. There's a centrifuge (price £3) for inducing screaming, vomiting and altered states of consciousness in children. At the Emporium it's all antique, retro and collectable, and next door an Oreo milkshake is £3.50, the specials are on the board, and you can 'compliment [sic] your meal' by bringing your own wine until 9 p.m. (excluding Sundays), and there's no corkage charged. In the curry house up the hill they do the Railway Beef Special As Served in First Class in Indian Railways, which, in the hill-station bungalows, the strimmers and car-waxers like to think is their natural diet. The Capodimonte's on steroids and the well-tuned lawnmowers hum like worker bees, and in the hairdresser's every one of the space-helmet hairdryers is occupied like on *Star Trek*.

I think I found Philpot here, standing on the beach in his long grey shorts, looking towards Belgium and asking whether the land had ended or the sea begun, because it surely couldn't be both;[16] trying to make out the words in the drawl of the tide; wondering why there were trains in and out if Herne was enough, as it seemed to be; watching horses pull coffin-carts to the graveyard and thinking that the departure of a person was like the passage of a pool of light on the sea beyond the pier. He knew the sour taste of despair, and took it as a sign that big, good things might have happened. He smelt Empire, as I smelt it at the curry house, and knew that across the sea there were others looking back at him. From the tenements he heard gin-powered blows landing in women's faces and learned that chemicals could unlock the doors to different compartments of a person, and he wondered how many compartments there were, and if there were other keys.

As a homosexual in the first half of the twentieth century, there were few places he could be himself. He straddled the edges of society, identity and the law, never knowing when the policeman or blackmailer would knock; never sure what the Church, which was his guarantee of eternal security, would make of him if it knew the truth.

The Renaissance art historian Giorgio Vasari is sometimes said to have held that every portrait is a self-portrait. It is true for Philpot. He depicts the dignity of the margins; the specialness of the ordinary. He gives to people of colour, to rag-tag circus performers, to down-at-heel workers, to anonymous nudes stripped of all identifying tags but their own wrinkles, the same care and reverence he gives to a dowager in a sapphire tiara. His figures are often at the seams of youth and age or youth and death; always in process, evanescent, as he was.

He painted two models repeatedly. Henry Thomas, his black servant and friend, and George Bridgeman. Were they lovers? We don't know, and to my eye he pays them no unusual type of attention. I suspect they were indeed lovers, and that he brought to all his subjects the same intense moral seriousness he felt for true intimates.

In everything he does he hints that his skill and his medium are not up to the subject – this despite great confidence in his own artistic technique. He gropes always over the edge of the immediately visible. One striking portrait of Thomas, for instance, has Indonesian batik as the backdrop, telling of an impenetrable cultural hinterland. This is not just a London servant, but a monumental and unfathomable story.

Hence, too, the mythological tropes that caught my eye that rainy afternoon in the Tate. All human stories are too vast and interconnected to be told without reference to an ancient and resonant *mythos*. All human lives are operatic. Philpot has the same sense of the interpenetration of this world and another that G. K. Chesterton and Charles Williams have, but without their hope. The mythic figures of which Philpot was particularly fond were fauns: animal–human hybrids. Often in his pictures it seems that the fauns are intended to *expound* the humans: to show what they really are. It is too obvious, and a slight to a sensibility as acute as Philpot's, to see them as mere proxies for a crude statement about sexual ambivalence.[17] It is not too crude to see them as statements that we all, all the time, flip between different ways of being.

'It seems to me,' wrote T. S. Eliot, 'that beyond the nameable, classifiable emotions and motives of our conscious life when directed towards action – the part of life which prose drama is wholly adequate to express – there is a fringe of indefinite extent, of feeling which we can only detect, so to speak, out of the corner of the eye and can never completely focus; of feeling of which we are only aware in a kind of temporary detachment from action . . . '[18]

Poetry, unlike prose, can operate in this fringe. Poets need peripheral vision. Poetry is representation of what we see at the edge of our normal sight. That must be poetry's main justification. Why should readers struggle to puzzle out the meaning of a poem if a few strands of prose would do the same job equally well?

Eliot intended his observation as a metaphor. We tend to think that metaphors are a way of *representing* the real world, but metaphors, if they are good ones, tend to be literally true too. There is no bright line between metaphorical truth and 'real' truth.

When I was starving in that Derbyshire wood, I saw at the edge of my vision and smelt on the edge of my olfaction and heard on the edge of my hearing things that weren't there. They couldn't have been. As I continued to watch, sniff, listen and starve, the edges swivelled to the centre of my attention, and the things that weren't there became more solid, smellier and louder than the things that were. I don't know what would have happened if I'd had the nerve to stay longer.

We must return briefly to Iain McGilchrist, whom we met in a slightly different context in Chapter 10, and whose brother we have just met on the shore of seventh-century Greece.

Consider a bird. It is pecking at grains of corn on the ground. To pick up each grain demands narrow, focused attention. That attention is provided, in birds and in us, by the left hemisphere of the brain. But that type of attention is not enough, for there is a fox crouching in the nearby bush. Unless the bird pays attention to the wider context, it will not notice the fox, and it will not go on pecking for long. That broader, more holistic attention – a very different kind from the narrow kind – is provided, in birds and in us, by the right hemisphere. If there is a healthy balance between the left and the right, the bird gets fed and lives to feed another day. If there is not, bad things happen.

McGilchrist argues that over the last three thousand years or so (but particularly over the last three hundred), the balance has gone wrong, and that accordingly bad things have happened and, unless the balance can be restored, will continue to happen. The left hemisphere has arrogated the function of the right, without any qualifications for the job.

The left hemisphere is conservative, nerdish, literal, loves generalizations, abstractions and inflexible categories, and confuses filing with understanding and intuition with white noise. It is meant to be an administrator, not a strategist. It controls the right hand, which, in most humans, is the hand which grasps and manipulates things. The left prefers maps to the hills the map represents. Indeed, its self-satisfaction is so complete that it thinks the maps – *its* maps – are more real than the hills themselves. It's a seductive and deadly heresy, as the wise know. 'The map is not the territory,' declared the Polish-American scholar Alfred Korzybri;[19] 'All models are wrong, but some are useful,' says a motto cited by wise statisticians; 'The menu is not the meal,' Alan Watts may or may not have said. Watts' formulation shows clearly the danger of the heresy, for if you eat only the menu you'll starve to death.

The right hemisphere takes the first look at the world and presents it to the left. The left takes that look apart. (If the right hemisphere malfunctions, the world is perceived as a disarticulated mass of parts. To the left hemisphere, a person is not a person but a right leg, a left leg, a pair of ovaries, and so on.) The left should then *re*-present its work to the right, which puts things together again for an integrated view of reality. The problem comes if the presentation doesn't happen: if the left hangs on to its dissected account of reality, insisting that that's how things really are. We murder to dissect, and we will have murdered the truth about the world.[20]

Unfortunately, as well as governing the manipulating right hand, the left hemisphere also rules much of our language, and so can give loud and superficially convincing explanations for its delinquency. If your right hemisphere is already debilitated, you might be taken in. Most of the twenty-first century has been.

The upshot is that we live in fragmented, deconstructed, uninte-grated universes. In maps rather than real landscapes. In worlds where everything is explicit. This is psychopathology. It is also demeaning and dull.

If we still have any right-hemispherical function we know that the really crucial things are not explicit: that music is not just the notes on the page; that to pull apart a poem in search of its 'real' meaning is to destroy it; that intuition is often a more reliable guide to conduct than utilitarian calculus; that love is more than chemi-cals and the chance of embryogenesis; and that metaphors are not only scalpels for the vivisection of life or engines for communicating ideas, but well up from deep within reality itself, and have some fun-damental concordance with it.

It is the right hemisphere that gives us peripheral vision, and which, for the same reason, understands poetry. Eliot was neurologic-ally correct.

If the right hemisphere's view of reality is the more correct one, and this view acknowledges, as it appears to do, that metaphors rep-resent something real about the cosmos, then I feel less uneasy than I did about my own metaphorical references to edges. It is right to cram into the same mental space, and the same book, mountain ledges, final breaths, the boundaries crossed by hunter–gatherers, tidelines, thresholds, solstices, and the frontier between right and wrong.

One final observation about the hemisphere hypothesis. It might be said that the left hemisphere is the hemisphere of edges. It loves divisions, after all. It creates them. Think of those filing cabinets. It's an edge-multiplier. The right hemisphere, though it sees the periphery, is an integrator. It puts the pieces together. Doesn't that mean that it is edge-phobic?

Not at all. The left hemisphere's divisions are dangerous because they are the hemisphere's own creatures, and do not correspond to the way the world really is. A human body does indeed contain a heart and two kidneys, but if we separate them the body dies, and so do the heart and kidneys. In the real world they exist as a unit. The right hemisphere, perceiving the relationship between the organs, acknowledges both the unit and the contribution to it of the individual organs.

The right hemisphere's holism doesn't mean that it is blind to edges. To acknowledge relationality one has to acknowledge the entities that relate to one another. This the right hemisphere does. It sees the frontiers and thresholds. It also sees the world of which they are a part. It knows that 'one' and 'many' depend on one another. The crudely dualist left hemisphere sees them as mutually exclusive.

The centre doesn't like this kind of talk at all. The multiplication of viewpoints threatens its status as the one true arbiter. The centre resents the many, for it believes that there is one, and that one is *it*.

If lured into epistemology, the centre tries a hackneyed gambit. You need a fixed point, it argues. Without a fixed point you have no idea where you are. And that fixed point happens to be the centre of the centre. The 'fixed' part of this argument is a superannuated Newtonian mistake. There aren't any fixed points. Everything, as Bohr saw, depends on the observer. (Blake, anticipating Bohr by a century, declared that 'A fool sees not the same tree as a wise man' and 'the eye, altering, alters all . . .')[21]

And what about the other part of the centre's argument: that the only sensible reference point is the centre of the centre?

It seems so reasonable. But it is egregiously wrong. So wrong that it deserves a chapter of its own. That's Chapter 14.

Now, though, we must look more closely at something we've touched on already: the idea that the cosmos itself is gradually unfolding; that new edges are constantly being revealed; that being alive is about surfing on the cusp of a breaking wave of reality, with newness constantly being revealed beneath the surfboard, whether or not we have the nerve to look down and watch it; that we join everything there is in the adventure of *process* – in a Great Procession.

12

The Great Procession

This grave contains all that was mortal, of a young English poet, who, on his death bed, in the bitterness of his heart, at the malicious power of his enemies, desired these words to be engraven on his tomb stone: 'Here lies one whose name was writ in water.'

INSCRIPTION ON THE GRAVE OF JOHN KEATS,
cimitero acattolico, Rome

FAR AND AWAY THE hardest event at the village fête was the slow-cycling competition. After tea and buns, we were each given a bike and had to ride along our own narrow lane on the school running track as slowly as we could. The winner was the one who arrived last at the finishing line. A foot on the floor meant disqualification. If you stopped pedalling or pedalled too slowly you fell off. If you pedalled too fast you stood no chance of winning.

The race was started and adjudicated by the buck-toothed vicar, brandishing his starting pistol and clipboard. 'It's all a metaphor,' he giggled. I think I know what he meant.

There had been a scene in the church the previous Sunday. The vicar had pointed to the plumber from the pulpit and screamed, 'You're an amphibian!' The plumber, baffled, but not sufficiently baffled to be un-offended, had stalked red-faced out of church, muttering that the vicar was insane and that he was going to resign as churchwarden. But again, I think I know what the vicar meant.

We're not *quite* amphibious. We live in time, but we are always seeking to crawl out of it – into belief in eternal life, historical biography, coach trips to castles, séances, family photo albums, junk shops, art galleries, the futures of our children and our species, next year's financial outlook, our own childhoods, and next week's weather. But even the relatively successful crawlers – I know a woman who barely leaves medieval Iceland, and another who inhabits a matrix of her own creation, set half a millennium from now – sometimes have the twenty-first-century tax man tap them on the shoulder, and one day they'll open the door to the Grim Reaper.

We can't reconcile ourselves to time being our natural habitat. Our language betrays our unease: 'Doesn't time fly?' 'Is that the time?' 'It doesn't seem a moment since she was in nappies.'

This, on the face of it, is rather odd. It has led in many religious traditions to the suggestion that we're not designed for time at all, but for eternity. Yet few of us seek death. Most of us, while trying to live in other times, try to extend our stay on this temporal plane. We want to be in *time*, if not in *this* particular time. We're constantly and inconsistently flopping over into the past and the future and back again, not quite at home anywhere, living, if you average it out, on the boundary between the past and the future – a boundary that isn't, for most of us, exactly the present. Living in the moment is an advanced skill, requiring years of arduous training in an icy meditation hall.

Humans 'give birth astride the grave,' pronounced Beckett, as we all knew anyway.[1] From our first breath we're diving down into the earth.

Beckett's crude metaphor doesn't do justice to our predicament. Both personally and as a species, we are Human Beings only because we are Human Becomings. Just what I am personally becoming remains to be seen. I have a few limited aspirations. But even on a biological level, it is clear that being and becoming are interdependent. I can talk about myself as a being only because I am also a process. If my metabolism stops its busyness, I stop my being. We've got to keep pedalling to make any headway.

As an undergraduate, I spent months trying pretty vainly to learn about cell membranes. That was how the physiology course started. The details were always fuzzy, but the general picture was that I was composed of trillions of cells. They were many, but they were also one (the one being 'me'). They communicated with each other via gates through their boundaries. Ions flowed in and out, and if the flow stopped, so did I. Such permanence as I had was dependent on the flow, just as the existence of the Heraclitan river, into which no one can step twice, depends on flow. Rivers exist: but what are they? *When* are they? Making sense of my biological existence meant realizing (though I never put it in these terms then) that stasis depended on flux, flux depended on stasis, and both stasis and flux depended on commerce across edges. I am a process dependent on being, and a being dependent on both processes and process.

Death, which shows that, however weirdly time might behave, the clock has been ticking and its hands inching on, is always an alien. Though death is the great fact, framing all our glories, we rarely acknowledge it. We assume it will happen to others, but not to us or ours. It is always a shock; an invasion; an obscenity. Though it is one of the two great defining edges – the other being our birth – and though we acclimatize to many scary edges over the course of our white-knuckle lifetimes – we never get acclimatized either to our own deaths or to death in general, though there are various techniques to help us to do so.

Decent Roman generals, on their victory parades, had a slave whisper into their ears, '*Memento mori.*' The old catechism advised Catholics, when retiring to bed for the night, to compose themselves as if for death. In some Tibetan traditions human skulls are used as drinking bowls. We might wear around our neck a small model of the Roman execution device used to kill Jesus (rather like having a golden electric chair on a chain), or have in our pocket a laminated picture of our favourite martyr. We might graft our mortality into our figures of speech, saying, as they do in Ireland, 'See you next week, if I'm spared,' or as in the Muslim world, 'I will be there, *Insha'Allah*' – if God wills it – and He might not wish me to survive that long. Or we might sit in the lotus position until our legs are numb, emptying our minds of vanities, seeking pre-mortem annihilation of the ego.

But these strategies are unlikely to recalibrate our intuitions. We recoil from dead animals unless they are smothered with gravy or remoulded inside a sesame bun, and we cannot see maggots as simply part of a benign cycle. The great ecologist E. O. Wilson declared that when he died he would like his body to be placed in a rainforest to be eaten by insects which themselves would be recycled. If that was not a pose, a slogan or a perversion it was brave and aberrant. We sympathize with Orthodox Jewry, for whom a human corpse is the 'Father of fathers of all uncleanness', and someone who touches a corpse a 'Father of uncleanness', who contaminates and excludes from the (now extinct) Court of the Israelites in the Jerusalem Temple anyone he touches.[2] In Judaism, though the corpse is in some ritual senses unclean, it is also precious. It must not be left unattended until burial. It must be washed in the ritual bath, or have water poured over it. Its nails must be cleaned. That too accords with our intuitions about dead bodies. When David Fuller, an English mortuary worker, was convicted of having sex with corpses, the outrage was greater than if he had committed rape.[3] We have a greater sense of the dignity of the dead than of the living. People can apparently cope with the idea of children starving in refugee camps, but not, as the Alder Hey scandal showed, with their hearts being stored in bottles.

The sound of mourning is not a grunt of ecological resignation, but a wail. Ululation should be taught in primary schools. Death, the most natural event, is the most unnatural. We're sure that we're not meant to die. A world with our death in it is a world that's all wrong.

In a Greek folk song, 'The Keys To The Underworld', a girl is given a glimpse into the underworld. She wishes she had never looked:

> *I saw the young men weaponless, the girls without their braids;*
> *I saw the darling little children rotting like withered apples;*
> *I saw the good housewives like doors torn off their hinges.*[4]

Though the singers of that song go diligently to the Greek Orthodox church every Sunday, their register is that of the *Iliad*, and the *Iliad* belongs to us all.

Our stories began when the edge of our mother collided with the edge of our father on a dance floor, at a lecture, or in an internet chat-room; continued when the edge of a sperm penetrated the edge of an egg; opened a dramatic new chapter when, with a scream that pre-figured the funeral, we were propelled over the edge of our mother; and will end, or at least acquire a massive suite of new adjectives, when we pass over the edge into the void. The euphemisms 'passed on', 'passed over' and 'passed away' reveal our underlying belief in process – a belief concordant with the way things are, and discordant with many of our systems of thinking. When Martin Luther bellowed 'Here I *stand*', he condemned Protestantism to a place alongside all other fundamentalisms, at odds with the constitution of the cosmos. For nothing ever *stands*, and any philosophy suggesting otherwise is futile.

Between birth and death it's a continual procession of edges; a procession maintained by the commuting of ions across the internal edges of the archipelago we call ourself, but defined by the loom-ing edge.

'I just wish life would leave me alone for a bit', said my friend Kate, after her latest scan. 'I'm seasick with all the ups and downs. I'd like to be asking for anything other than a bucket.'

We're swimming in a rough sea, with waves lifting us up and slam-ming us down all the time. It makes us queasy, but perhaps the waves are carrying us on; perhaps we would sink if they were not there. But whether they're carrying us or not, we're crossing their crests, and the crests give the sea its only form, for there are no other landmarks. But there is a roar, audible all the time if we listen for it, and sometimes deafening, which tells us that there are rocks ahead which we will reach when the other edges are passed, and which will remake us.

Though we fear change, we are dissatisfied without it. When we are ten we want to be twenty, and when we are fifty we want to be twenty. Adult birthdays, beyond the age of thirty, are almost always dismal occasions which have to be sweetened with cake or palliated with wine. When we are poor we want to be rich, and when we are rich we want to be richer. If we are John we want to be Tom or Joanna. We hate our governments, but as soon as we have a new one, we hate that one too. We want more sex or more holiness or, more likely, more holiness in our sex; we want the desk with the view of a tree, the latest incarnation of the iPhone, a boyfriend who prefers *Fidelio* to football, or a girlfriend who loves us rather than the idea of marriage. We look forward to retirement (though not to the incontinence that comes with it, or the extinction just the other side).

We're perpetual discontents, wanting always to cross over, to be elsewhere. As individuals and societies, we want always what we can't have. A strip of fly-blown land bordering our own. A job for which we're not qualified or for which we're overqualified. When Venice was just mudflats it lusted after rolling hills. It then acquired the Veneto, got its hills, celebrated them for a while in a form of unfashionable pastoral painting that everyone else rather despised, and then got bored with pastoral and reverted to its ancient obsession with the ultimately inaccessible – the transcendent, glittering God of Byzantium, seen through a Catholic lens.

Sometimes dissatisfaction has been turned into an art form as, for instance, in the ideal of courtly love. Those knights need not have been lonely. But they chose to give their love exclusively and hopelessly to the inaccessible wife of another. Unreachability was the point.

Sometimes it has been an *actual* art form: thus Mondrian's pure blocks of colour, which say that perfection, at least of red or yellow or squarishness, is possible, but only within the grid of the picture. Move outside the grid and you'll be back in the messiness of the gallery and your tangled life. The gulf between that life and the grid is unbridgeable. Mondrian says that in the real world outside his grid only shabby compromise is possible.

There is a whole religion devoted to looking forward. It has its own creed: *progress* – the idea, supported only by a very select-ive view of history, that some sort of dialectical ordinance decrees that things can only improve. It has its high priests – Steven Pinker is the highest – who try to persuade the faithful to keep their chins up in the face of climate change, war, the meltdown of democracy, the tsunami of meaninglessness and alienation, the polarization of political discourse, the torching of high culture, and the horrors of daytime TV. These are blips, say their sermons. The general trajectory is onwards and upwards.

The appeal of such entreaties is obvious. We and our children are processing into the future, and we would like to believe that the promised land is over the horizon, even if we won't enter it our-selves. But even when this is impossible, our appetite for the future is insatiable. Most of the broadcast and print media is unapologetic prophecy, delivered with a confidence made possible only by the knowledge that human memories are dismal.

But whether we're straining our necks to see into the sunlit uplands, or bending them to gaze into the slough of despond, what-ever is over the edge of the *now* is consummately interesting.

More interesting than trying to see the future are our attempts to clamber back over the edge of the present and inhabit the past. We all do it, whether by living where we've been brought up, or looking through the family photos, or hugging the childhood teddy bear, or reliving our childhoods through those of our unfortunate children. We are all, in our incessant history-telling, obsessive autobiographers. We feel a need to re-write the past – to make it better than it was, or more exciting, or more consistent with what we have become, or to create a personal myth of a personal Golden Age.

All communities and individuals have their Golden Ages. Yet it is not usually our past per se which so fascinates, but our childhood. Why then? Why are our efforts to get back to being a student no more determined or systematic than going to a 1980s disco, whereas we have evolved elaborate philosophies to explain our devotion to childhood?

Only by becoming a child – indeed, only by being born again – could anyone enter the Kingdom of Heaven, said Jesus.[5] Thomas Traherne wrote that in his childhood, '... All appeared new, and strange at first, inexpressibly rare and delightful and beautiful ... My knowledge was Divine. I knew by intuition those things which since my Apostasy, I collected again by the highest reason.' His ignorance was a gift, not a disability. 'Eternity was manifest in the Light of the Day, and something infinite behind everything appeared ... The city seemed to stand in Eden, or to be built in Heaven ... all the World was mine; and I the only spectator and enjoyer of it.'[6] Wordsworth and the other Romantics followed suit. For them, childhood was a time when everything was new, and therefore viewed with eyes unsullied by presumption, and therefore seen accurately.[7] Picasso agreed, declaring that he had spent his life trying to paint like a five-year-old, because only then would he be painting what was really there. These epistemological justifications draw on Plato, who held that all true knowing is *anamnesis* – unforgetting. We want to be children again, the reasoning goes, because only by doing so can we unforget our mistaken conclusions about how the world is, and thus know how to live.

These are powerful ideas. But, I suggest, they are secondary to a more fundamental explanation.

Childhood is the time before we cross the catastrophic edge between childhood and adulthood. We can theologize the transition, if we please, as a fall: as the biting of an apple; as the acquisition of debilitating knowledge. That's a helpful way of looking at it, so long as we realize what the fatal knowledge is. It is the recognition that there *are* edges, and that life hereafter is going to be characterized by them. It resolves soon, and terribly, into a knowledge of mortality, but at its inception, and most basically, it is a recognition of *process*: of churning, moulting, recombination and moulding. Until then, we have occupied an immutable present. But now we must begin to breast those waves, which is exhausting – and realize, too, that what we believed about ourselves, and about this place, was incomplete: more accurate in some ways; less accurate in others. The new life is lonelier and more strenuous. It takes some getting used to. Most of us never manage.

Some do, mostly, as we've begun to see, by living consciously on the edge until some of the vertigo starts to ebb. There are many ways of doing that. One is constant physical movement, like those wandering hunter–gatherers who I've suggested are the normative humans. Another – a sort of metaphorized hunting and gathering and walking – is artistic or scientific exploration, in which ideas are tracked to their source and speared; moments collected and stored on a canvas for when the weather turns.

The more direct the confrontation of the edges of time and process, the more fecund the art and the science. Like Whitehead (whom we met in Chapter 9), I hate the term 'great art' (it has unpleasant centrist connotations), but if it means anything, surely it means this. It connotes not just a dextrous brush but a clear view of the edges, a determination to face them down, and an angular – and therefore bearable – final view of the world before the dark closes in. If that's done bravely, the edges begin to look less fearful, and the picture (or book, or poem, or whatever), as it demands a response from us, begins to elicit the response: 'I see. Time passes. And it is all right.' It is the sound of our own voice saying this that really stills us. Even if we're in dialogue with a towering genius, that sound will take us back to the safe place of childhood, before edges began to growl.

I wouldn't scoff if you said that Beatrix Potter worked for you. The Zen masters are just *too* direct for me: they don't have that quality of angularity, necessary for turning the terrifying immediacy of their vision of the edges into something I can use. For me the supreme edge-artistry is found in Rembrandt's self-portraits. He is the man of all genres and none, who rides over all edges and drains them of their terror. His smiles are all half-smiles because he's seen the meaning of things, and knows it is partly a joke and partly deadly serious, and he knows too that if you keep your nerve there's nothing to be scared of.

Whatever method you use to break the power of the edge by being mindfully on it, the ultimate object is to be free: 'So long as he kept moving he would be all right,' wrote Peter Matthiessen of a character in one of his novels. 'For a man like himself the ends of the earth had this great allure: that one was never asked about a past or future but could live as freely as an animal, close to the gut, and day by day by day.'[8]

If we're in a world of frontier-crossing – and whether we're nauseated or exhilarated by the crossing, and whether or not time behaves politely – it helps to have some signposts; to know something of the geography.

13

Marking the Edges

*... Man is a Noble Animal, splendid in Ashes, and pompous in
the Grave, solemnizing Nativities and Deaths with equal lustre, nor
omitting Ceremonies of bravery in the infamy of his Nature ...*

SIR THOMAS BROWNE,
Urn Burial[1]

ONE OF THE ONLY real feathers in my cap (apart from my superb
uncoolness) is my membership of a raggedy, disreputable,
anarchic folk band that plays just twice a year: on May Morning and
on the Day of the Dead, All Souls' Night – pivotal edge-times.
We call ourselves the Whirly Band – though the real name is the
Hurly-Burly-by-God-It's-Early-We're-the-Whirlies-We've-Got-You-by-
the-Short-and-Curlies Band.

We get up before dawn on 1 May and start to cross frontiers as
soon as we dress, for we dress all in green, and cover ourselves with
foliage. Then we head off to the middle of Oxford and subvert it,
and take possession of one of the most central places there, just next
to the august stone heads by the Sheldonian Theatre at which the
Obscure Jude gazed when he was starting to sense that he was a perpetual outsider and would never storm Oxford's ramparts.

Much of Oxford has been revelling that night, and quite right
too: it's wise to be wakeful when a new age dawns. But everyone,
however dissipate and liverish, stands respectfully to listen to the six
o'clock songs from the tower of Magdalen College. As the last note
fades, they bawl raucous approval up at the tower and stream up High
Street and on to Broad Street, for that's where the real business of

May Morning is, and that's where we are. We can hear them coming. They're like migrating wildebeest, and their Gucci-shod hoofs ring loud on the cobbles of Radcliffe Square.

As the first wave turns the corner, our piper fills his bag with air full of hawthorn and beer and vomit and the breath of the newly arrived swallows, stamps his foot to mark the time, the deep bass drum takes it up, and we're off.

The tunes are ancient, incantatory and most are very simple. Many of them are Basque, Breton or French. When they're French they're from dripping homesteads in the Auvergne. When they're Basque or Breton they're from bitter headlands or from proscribed cider sheds and cellars. They're emphatically not the thing the stampeding wildebeest like. They plan to migrate next to the London offices of merchant banks and brokers. There they will continue to travel in herds from tip to tip, pub to pub, fashion to fashion, market peak to market trough, and though we don't think of wildebeest as regal animals, they'll rule us if we let them.

The tunes are not just incantatory, but mind-altering and shapeshifting. The transformation often happens very fast. I never cease to wonder at it. The wildebeest shed their hind hoofs. Human feet – which are dancing feet – sprout in their place. They shed their fore hoofs too, and human hands – which are for waving, clapping and stroking – appear. The wildebeest also shed just about everything else – at least while the music lasts. They join their new hands with the hippie girls in tie-dye skirts and the farmers in dung-stained tweeds and the momentarily unselfconscious dons in terylene trousers, and they circle round, the strands of the dance weaving intuitively around one another, as if the dancers had been taught the patterns at their mothers' knee. (It's a textbook example, I can't help noting to myself, of the dissolution of social hierarchies laboriously chronicled by Victor Turner, the main architect of the idea of liminality.) No one's shouting instructions, but all the left feet hit the ground at the same time; the joined hands are raised symmetrically in a bridal arch for newly joined couples to pass through. The steps come straight out of the tunes, and if you're inside the tune, surrendered to it, you might as well try to stamp your foot out of time as stop a storm ruffling your hair.

'Romantic tosh,' said someone, when I told him this. 'Wishful projection,' said another. Not so. Come and see. It's not remotely supernatural. It's just a lot more natural than we're used to, and the truly natural is truly exciting.

They'll deny it in the pub that night, but those wildebeest are happier than they've ever been. Perhaps a fragment of memory will stop them, twenty years from now, investing in oil or arms or trying to become the party chairman. That's the sort of thought that makes me blow my tin whistle with evangelistic passion.

It's not the music that does it. The music is just an agent of the edge-time. It's the time, and the surrender to its edginess, that are doing the work. The wildebeest might not surrender to edginess again until they get their own terminal diagnosis.

'Liminal' is the most overused word in the modern lexicon, with the possible exception of 'unprecedented'. There's a good reason. We tend to talk about things that happen rather than about things that don't, and things only ever happen in liminal spaces: in transit lounges; when we don't know what we are because we're in the process of becoming; when we taste *process* itself – the *terroir* of the cosmos. If you're in process, you're continually crossing frontiers. 'Liminal' describes *everything*. Everywhere's a transit lounge.

I'm not talking about human life being a preparation for eternal life. I'm talking about human life being a preparation for human life. For Jungians, liminal spaces are where self-realization occurs.[2] But doesn't self-realization happen everywhere? Every moment, in every place? For Gawain, Wales is the borderland – a semi-familiar place through which he passes on the way to the Wilderness of Wirral where the action happens; where he finds out who he is. Everywhere is Wales. Every moment, in every place, we are becoming the creature who will cross the next boundary, and even the most unreflective know something of what's happening to them. Only in extreme and extremely perilous states do we have no idea. 'So full was I of sleep at the moment in which I abandoned the true way,' wrote Dante.[3]

Though everything's liminal, there are *special* liminal places and times.

Old woods, where trees spring from the corpses of trees, are, along with the sea, the labour ward, the care home and the cemetery, the places where *process* is most unavoidably on display. Trees, though fixed in the ground, move. It took me thirty-five years to lose my hair; trees can lose theirs in one stormy night. Our hair turns white over sixty years; trees go from green to gold and back again in one. If I lost a limb I'd see it as a major, life-defining event. Trees shiver off their limbs all the time. If you have a deciduous tree in your garden you can throw away the calendar: its colour marks the time. Trees show how time behaves and how things happen.

No wonder there are happenings in woods. Many Arthurian adventures start there, if the adventurer is fortunate enough to get lost. Gawain rides into a forest 'that was deep and fearsomely wild' with 'high hills at each hand, and hoar woods beneath of huge aged oaks by the hundred together; the hazel and the hawthorn were huddled and tangled with rough ragged moss around them trailing, with many birds bleakly on the bare twigs sitting, that piteously pipes there for pain of the cold.'[4] As we've seen already, he is so thoroughly lost that even the *names* of the things around him evaporate.

In C. S. Lewis's *The Magician's Nephew*, the 'wood between the worlds' is just that – a departure and arrival lounge through which one has to pass en route between other realities.[5] Dante, like those Arthurian knights, was lost in a dark wood before his adventures in Hell, Purgatory and Paradise.[6] The druids sacrificed in their sacred groves; Robert Graves[7] and James George Frazer[8] banked their academic reputations on the notion of trees as portals to the numinous; modern hikers, swathed in Gore-Tex, get their own spiritual experiences by striding along forest paths; and the pillars of temples and the buttresses of great cathedrals are just stone trees, so that you can sacrifice in a sacred grove without getting wet. The stone efflorescences that are the capitals of Ionic and Corinthian columns declare loudly that you're in the forest. So does the Green Man who leers at Holy Communion in the Gothic cathedral.

Then there's the sea, always on the edge of our consciousness, and on the edge of my hearing as I write this. The sea where all life began. Our ancestors crawled out of it, and we will return when the rain leaches our molecules from the cemetery or the smoke cloud from the crematorium drops us. The sea that chews rocks and whales into sand, multiplying the number of edges with each grind of its molars. It is fearsome to think that there's a maw like that at the end of every road. The sea is patrolled by Leviathan, the ancient enemy of God. It is dark, cold and full of eyes. No wonder John of Patmos, looking out at the eastern Mediterranean and the end of all things, looked forward to the time when there would be 'no more sea.' Adventures here are for the adventurous indeed.

But since everywhere, for an edge-person, is an edge-place or a place between particularly tectonic edges, anywhere can become specially edgy or specially liminal. If we're tuned right, we can make anywhere one of the Celtic 'thin places'. Or, better: if we're tuned right, we'll know that everywhere is one of those places. You needn't go to a sacred grove or a stone altar still stained with the blood of a sacrificed slave, or crouch in the rearing bows of a boat ploughing its way to Patmos. You needn't shiver on a summit where, below you, you can see the sleeve of a half-buried climber who didn't make it and has been hunched there for forty years thinking of his wife and children. You needn't watch a storm bundling up birds and flinging them into the sea. You needn't rise in the dark in a jungle ashram to chant mantras until the monkeys in your head stop chattering and you can see what's left of you. You needn't run all day and night across burning sand until you are just thirst and shredded feet. You needn't stare into candles in a crypt.

The thin place might be the work station in the open-plan office where you got the email saying that your father was dead, or the one saying that you were probably going to die too, but that there was a new treatment that might be worth trying, and would you make an urgent appointment. It might be the bed in that Airbnb cottage in Provence where your child was conceived, or the bed in Birmingham where nothing at all could be conceived. It might be the stop where you catch the bus to the hospital, or where you once saw, but never spoke with, the woman who would have made all the difference. It might be your grandparents' houses, where just about everything happened that, long before you were born, determined what you became. It might be the caravan where you were abused, the cake shop where you first learned about marzipan, or the zoo where the twitching nose of a companionable anteater told you that you were a wild thing too.

Are these edges? They are, at any rate, defined by edges, and in turn make us, for good or ill.

We're told that space and time, if they exist at all, are bound together as indivisibly as the hypostatic union. If so, it is strange that we, who swim in time and shuffle through space, and whose cells have both dimensions and clocks, do not consistently feel this to be so.

Often, at particularly special, particularly liminal, particularly edgy times, space and place evaporate, taking with them our habits, pretensions, prejudices and expectations.

Is it fanciful to suggest that a particular time or chronological frontier should have a particular power? Perhaps – but that fancy is shared by all cultures and all ages.

Twilight – when day bleeds into night, transgressing the boundary between light and dark ordained in Genesis – is a more haunted time even than midnight (when one day becomes another) or midday (a perilous time in the mountains of the Peloponnese, when, at the cusp of morning and afternoon, Pan stalks the land, inducing *Pan-ic* in anyone unwise enough to sleep in the shadow of a tree). In the Bhagavata Purana, the demon Hiranyakashipu cannot be killed by either man or animal, with bare hands or with a weapon, inside or outside, on the ground or in the air, by day or by night. Yet this does not make him invulnerable. Vishnu transforms himself into a human–lion hybrid and with his claws kills Hiranyakashipu at dusk, on the threshold of a palace, as the demon squirms on the god's lap. In the Mahabharata, Indra, who has promised not to kill Vritra (a danava, who personifies drought) by daylight or dark, with anything dry or wet, smothers him with foam at dusk.[10]

Many religions have one special day a week, and many during the year, when the devout may believe their prayers have particular potency. Solstices, equinoxes and quarter days (when one season becomes another) have long been times of sacrifice, propitiation and fete. The winter solstice, for instance, is celebrated across the globe. In Egypt it was the feast day of Horus; in Greece the birthday of Dionysus, Ceres and Hercules. In the Roman Saturnalia, celebrated originally on 17 December, the social order was temporarily upended: slaves became masters and masters slaves. For Christian believers the shaman Jesus, born of both divine and human stock around the time of the solstice, is the only adept navigator in the space between life and death, and he upended more than just social order. In many countries, the first person to set foot over the threshold in the New Year is thought to determine the fortune of the household for the year to come.

If religion and folklore will not convince you of the significance of threshold times, a spouse's outrage at a forgotten birthday or wedding anniversary certainly should.

If we're beings in process, being transformed by our passage through liminal times and spaces, and it's a good idea to acknowledge what we are, then we'll mark our journeys. We'll see the transition between stages as significant. We'll have rites of passage.

It's interesting that the only rites of passage worthy of us – those marking birth, adolescence, mating and death – are either religious or parrot religious tropes. Mostly, with the eclipse of religion, rites of passage have become meaningless and degrading. Adolescence is marked not by reading the Torah to the critical congregation, or being scarified for a demanding god, but by drinking white cider in a bus shelter, losing your virginity behind the pub, or getting a new phone. When you die, your departure won't be marked by a wailing coven of black-robed women. Your relatives won't dig you up later, wash your bones in home-made wine, and store them in your family's house of the dead, as they do in Greece. Instead, you'll have flowers from the petrol station and 'Hey Jude' over the PA system before you're baked in a municipal oven on the other side of the ring road, your fumes making the downwind children cough and your carbon dioxide contributing to the mass extinction of insects.

However secular you think you are, I'd be surprised if you didn't agree that the religions do you more justice than John Lennon; that you signify more than that elaborately sombre council official with the cremation schedule on his clipboard has recognized. The most hard-nosed materialist handles his dad's ashes with philosophically inconsistent reverence. We need a tradition that does edges properly.

The anthropologist Victor Turner thought that liminality was unstable and unsustainable. Well, yes and no. We're all, all the time, in the liminal space between birth and death, and in lots of other liminal spaces too. In that sense, liminality is not only stable and sustainable, it's all there is. But liminality entails movement, and it propels. So staying in any particular liminal space is impossible.

It's a good idea to know that you're somewhere in these spaces, but if you're doing life properly, you won't be able to state your position. If you're travelling constantly, by the time you've finished saying where you are, you'll have moved on, and your statement will be wrong. The only accurate statement is: 'I'm on the move.'

Once I skied to the geographical North Pole. The Pole is an abstract point on the shifting ice of the Arctic Sea. I only knew I was there because the GPS said 90° 00' 00". But it said it only for a moment. I looked up from the GPS to savour the moment, and when I looked back I'd moved. Life's like that.

There's a sophisticated culture devoted to pretending we're not in transit; that we can state our ontological coordinates precisely; that we can know where we are. Many iterations of this culture are explicitly religious. Or they say they are. In fact they are anti-religious. They are the religious fundamentalisms – as dissonant with our real experience as the fundamentalisms of material reductionism. Most of us, at least in the early hours of the morning, when certainties are most fragile, aren't convinced by the pretence. We know that we don't know; we know that we're accelerating in the dark, towards the dark.

We are the religious, however malevolent or non-existent you think God to be.

As I write this I'm sitting alone, in the liminal space between Europe (whose soil I'm on) and Africa (whose air I'm breathing and whose sand fills my remaining hair and my eyebrows and stains the car); in the space between this moment and the next, this thought and the next, this lump of hazy self-consciousness and the next, this breath and the next, this heartbeat and the next.

I've spent a few thousand words suggesting that the liminal is where we all are; what everything is. Isn't this a lonely, wretched vision? If you do an internet search for images depicting liminal spaces, what comes up is desolate, modern and artificial. There's rarely a tree or a square of grass to be seen, and, most importantly, there's rarely anyone else there. Think of a deserted filling station at night, lit by clinical lights that banish shadow and depth.

The absence of anything natural underlines the crucial loneliness of liminal spaces as we, in our twisted, modern way, perceive them. If even a clump of moss were visible, we could feel companionable towards it, and would be deprived of the disorientation often said to be of the essence of liminality, which enables you to be disabused of your pretensions – cut adrift from your moorings so you can move on.

Turner suggested that liminality, of which a feeling of isolation is an indispensable ingredient, was central to the generation of fellow feeling. He was clear. He gave the example of St Francis. Francis, he said, 'appears quite deliberately to be compelling the friars to inhabit the fringes and interstices of the social structure of his time and to keep them in a permanently liminal state where ... the optimal conditions inhere for the realization of *communitas*.'[11] That sort of liminal state is very different from the one in the internet images. For liminality to do the sort of work envisaged by Francis, we need to rehabilitate our view of liminal spaces. A good liminal space is busy: it's populated by all kinds of entities, human and non-human. We look at them, we feel isolated from them, we don't like the isolation, we start working to reduce it, and we're transformed.

St Francis's strategy has been widely adopted. Presumably it has been widely adopted because – based on a better account of liminality than the one in those internet images – it works.

One example will make the point. The monks and priests of the Ethiopian Orthodox Church set themselves strenuously, if not downright brutally, apart from lay people, with the explicit aim of qualifying themselves as brokers of *communitas*. Priests fast for 252 days a year. Exactly what that fasting involves will vary, but some will be near the edge of survival. A would-be priest must spend two years as a wandering mendicant, far from family, friends and institutional charity, begging for everything. Some gravitate to Addis Ababa and sell lottery tickets. One priest told a British researcher that he remembered being 'perpetually hungry, being cursed and abused by people he begged from', and was terrified of being savaged by dogs. Monks, the researcher noted, were more effective *communitas*-generators than priests because monks mortified the flesh more diligently. Priests, for instance, can have wives; monks must be celibate.[12]

This way of life involves many of the aspects described in the Sermon on the Mount. But in the Sermon these aspects are supposed to make *you* blessed, whereas in the Ethiopian model they seem to make you better able *to* bless. Perhaps you can only be blessed yourself if you actively bless others? Here is an obvious parallel with the shamans we met earlier: they go to wild, distant places where the lay people cannot or will not go, and whatever they return with acts – by whatever psychological or theological mechanism – as some sort of social adhesive.

It is all rather counter-intuitive. You'd expect an emphasis on the margins to *diminish* the power of the community. How might it work?

I have two suggestions. First: to stare at the margins diminishes the power of the centre. Power in human communities is almost always centripetal – about increasing the power of the centre itself; the few. It is rarely concerned with bonding the people scattered throughout the community.

Second: if a community has in it someone who lives visibly on the margins, each individual is reminded, every time they see the threadbare, emaciated priest – the representative of the community – limping through the village, that there, on the margins, is where we all always *really* are. That generates a sense of community: we're all in it together; we're all poised in the space between heartbeats; and one day soon, each of us will topple over the edge to join the massive, aristocratic community of the dead.

Before we do that, though, there's a lot of exhilarating, edgy living to do. We'll live better if we can stay out of the centre's clutches. That's best done if we know a little more about it. So now we'll go briefly to the centre(s) of the Earth.

Hold your nose.

PART 4

A Journey to the Centre of the Earth

14

The Great Fraud

*A point is a 0-dimensional mathematical object ... Although the
notion of a point is intuitively rather clear, the mathematical machin-
ery used to deal with points and point-like objects can be surprisingly
slippery. This difficulty was encountered by none other than Euclid
himself who, in his* Elements, *gave the vague definition of a point as
'that which has no part'.*

https://mathworld.wolfram.com/Point.html

THIS MORNING I DROVE along a road that snakes slowly up a big
mountain above the sea. I know this road very well. Because of
its many hairpin bends, it has long sections where the gradient is not
particularly steep. But there is one place where, after a flattish stretch,
the road suddenly leaps into the sky.

The car has one of those devices that tells you when you should
change gear. It is based on the revs. As I approached the big jump,
I changed down to second gear, knowing I'd have to move to first a
moment later. The monitor protested, and urged me to go to fifth.
That would have been a very stupid thing to do, given what I knew
was ahead. The car didn't know the road as I did. The car's advice
was based on averages: on average, if your revs are such and such, you
should be in fifth. Fair enough. But there are no average roads. If you
assume there are, you'll stall.

Perhaps you think I've overstated my hypothesis about edges.

If so, here's another reason to take it seriously. As a matter of
statistics there's nothing but edges. Everything is eccentric. We have

only uncertainty. The orange yogi-man with the big beard was right to try to rehabilitate – or rather habilitate – doubt in the minds of those cross-legged schoolboys, certain of their place in society and eternity.

Even our certainties are founded on uncertainty. I breathe air in a room, confident it will be replaced by new molecules. But it is only *probable* that it will. The air molecules are bouncing off one another in wholly random Brownian motion. My side of the room could soon be wholly airless. It's unlikely, but possible. Our apparent certainties are based on mere probabilities; are produced by randomness; are creatures of uncertainty. The supposedly immutable laws of physics have not been decreed by some central legislator. There is plenty of room for God, but he, she or it does not legislate in the way of central government, but delegates, at least very often, to the autonomy of particles small and big; particles alone but in community; particles in tight and interdependent conglomeration.

When the centre insists on its right to rule, remind it (not respectfully) that, except in a merely geographical sense, *it doesn't exist*. Most non-geographical senses of the centre are statistical fictions, based on the notion of the average. And there simply are no average things. You've never seen, eaten or had lunch with an average. Averages are fabrications, sometimes convenient, and often sinister. Sometimes we have to kowtow to them. That might be one of the clauses in the social contract we've signed. But if we do kowtow we should do so reluctantly and suspiciously, knowing the price we're paying. If your doctor prescribes treatment based on the study of big cohorts, know that the pills are directed at a wholly fictional patient deemed to be at the centre of the cohort. It's ham-fisted, but pragmatic and possibly justifiable. Just be aware the doctor is not treating *you*. She says she's practising 'evidence-based medicine'. The evidence on which her medicine is based is evidence about the efficacy of the treatment in a wholly non-existent patient.

Outside science, the tyranny of the algorithm is more obviously hazardous, and resistance – given the ubiquity of algorithms – is increasingly hard. Everything done by AI is based on averages; everything done by AI necessarily misrepresents and demeans the individual.

We have seen already Nassim Taleb's demonstration of the woeful unreliability of predictions based on past events. But why is this so? Because, says Taleb, '[a]lmost everything about social life focuses on the "normal", particularly with "bell curve" methods of inference that tell you close to nothing. Why? Because the bell curve ignores large deviations, cannot handle them, yet makes us confident that we have tamed uncertainty. Its nickname in [Taleb's] book is GIF, Great Intellectual Fraud.'[1]

The fraud is everywhere. We all fall for it. It fuels a perverted appetite for data, on the fallacious ground that the more information we feed into the bell curve the more accurate its predictions will be. If the model is defective, no amount of information is going to make its outputs any more reliable. More data will just increase our misplaced confidence in the GIF.

The fraud, as we noted in Chapters 4, 5 and 11, justifies top-down theorizing and edict. We submit to the edicts because of our faith in the model; because we are told that they are an irresistible inference from the way the world is. But the world *isn't* like that. In the realm of causation it is a world of black swans. Many of those black swans are human. Humans don't go by the statistical manual. Taleb again: 'I don't particularly care about the usual. If you want to get an idea of a friend's temperament, ethics, and personal elegance, you need to look at him under the tests of severe circumstances, not under the regular rosy glow of daily life. Can you assess the danger a criminal poses by examining only what he does on an *ordinary* day?'[2]

Here's my gloss: to find out what humans are, and to draw worthwhile conclusions about their behaviour, study them in their natural habitat. Studying animals in their natural habitat is a surprisingly recent development in the study of animal behaviour. It had long been presumed that we could know everything about rat behaviour from observing rats in the laboratory. It is not so.[3]

Even if the bell-curve model were robustly predictive, it would still be morally and politically necessary to resist its conclusions because of its totalitarian corollaries.[4] It treats humans as figures rather than, as they truly are, stories. It makes horrors possible. All dictators know it. 'One death is a tragedy', said Stalin, with his homicidal cynicism. 'A million is a statistic.' There's nothing offensive about sending a statistic, or even a million statistics, to the gulag. Abstractions don't have souls. Abstractions are also, note, the main currency of the left hemisphere. 'One can't love humanity', said Graham Greene. 'One can only love people.'[5] That's the right hemisphere's response to Stalin.

States and corporations don't exist. They are wraiths, made of mist and the fantasies of lawyers and theorists and men (they're usually men), who use them as platforms and milch-cows. 'There is no such thing as society', Margaret Thatcher is reported to have said (she didn't, actually, but she certainly believed it).[6] She was wrong about almost everything, but for once she was on to something, though she disastrously misunderstood the implication of what she was saying. There are glorious conglomerations of individuals, in which individuals can thrive in a way that they cannot alone. But the conglomerations themselves do not deserve names. To identify them, you'll have to list all the constituent individuals. Sometimes, for a while, a wraith might help a red-blooded human to be more fulfilled. If so, good, but never forget that it is a ghost in service to a *real* thing: a fantasy that might facilitate a true story.

'But I do have a centre', protested my friend Penny. 'It's real and comforting. It's where I am. It's where I'm from. It's where my stuff is and where my friends are. It happens to be in the middle of London too, which I suppose makes it doubly dubious to you. It matters very much to me. And now you're telling me that it's not there at all; that in being happy and settled in my humble little inner-city flat I'm the victim of some great confidence trick. It doesn't feel like that to me. Did it feel like that to you when you came for dinner last month?'

It didn't feel like that, and the reason is that however virtual much of our life is, where we live is a real place. A physical place. Such places, *simply by being physical*, are a repudiation of centrism, which is not a place but a theory; a notion, which can therefore accommodate only purely notional people. Real, non-notional people live in places which they might call their own centres, but which are very different from what the centre calls a centre. These real living places are houses and families and communities. A centre is an infinitely small point on a map. Penny's flat is not infinitely small. It would never fit in the centre.

When humans fall in with the centre's insistence that it is real, and should be treated as more solid than humans themselves, tragedies happen. Sometimes they are the Stalinist type of tragedy. More usually they are the corporate minion type.

We've all seen it. I recently sweated in a hot office for five hours, waiting for a stamp on a piece of paper, watching fans jerk around without stirring anything, looking at clerks sitting shamelessly at empty desks, their hands folded unapologetically on their laps, making no pretence to themselves or anyone else that their work was useful or significant, and at piles of untouched and mouldering paper on the shelves, gnawed by termites. The clerks weren't trying to be obstructive. For them, their work was all about the choreography of deference and power. For most, it wasn't about ambition. Ambition at least entails motion. This was a static ballet; it was about holding a pose.

They weren't nasty people. But they endangered others and, most obviously, themselves. If you hold a pose like this – a pose of genuflection to the centre – you'll find in the end that you can't unbend to hug your children; that the passive bureaucratic face can no longer curve in delight, anger or outrage.

In *The Satrapy* (see p. 58), Cavafy depicts a life corroded by compliance and sycophancy. The corrosion is so severe that even the desire for better things is corrupted. The satrapy does not desire the 'other things' (presumably including the joys of philosophy and art) for their own sake, but for 'laurel wreaths' and the 'acclaim of City'.

This is what the centre – the static point – can do to individuals. The State's centre becomes *their* centre. It's a process of colonization like that of the ichneumon wasp that consumes the body of a caterpillar from within, or the fungi that colonize and direct the brains of ants.

One of the most pleasing ways to deploy these thoughts is in arguments with your teenage children, whose desire for the anointed trainers or phone of the moment is a desire to be inside, and at the peak of, that bell curve. 'You know,' you can say, 'you're trying to be something that doesn't exist. You're aspiring to be the average customer. There's no such thing. Do you really want to unwish yourself?'

It won't change the children, but it might make you feel better. And if your children are still in the room you can indulge yourself further by reminding them of the gratifying language of *regression* to the mean – one of the axioms of genetics, which says that offspring are likely to be nearer to the centre of the bell curve than their parents. 'Surely you don't want to *re-gress*, Jonny? Don't you want to move away from the centre, to the edges of the bell curve and to the thrilling bad lands beyond, where there be dragons?' (The answer to the second question, whether delivered in words, body language or by contemptuous exit, is always an emphatic 'No.')

How were children persuaded to use their backs and chests as advertising hoardings, bearing the single word *Adidas*? And to pay Adidas for the privilege of advertising their wares? It doubtless has something to do with children's roots being cut, and them feeling the need to go off in search of an identity. But *Adidas* as an identity? That's weird. It takes serious centripetal force (and over €2.5 billion a year in marketing and point-of-sale expenditure[7]) to produce a distortion in the human psyche as profound as that.

How far should we carry our suspicion of statistics, models and abstracts? Does acknowledging that the world of causation doesn't function like a casino mean we should declare that all real swans are black, and that if you think you've seen a white one you are deluded?

A full answer to these questions is well beyond the scope of this book. But when taking account of uncertainties that are dependent on humans, it is probably prudent to disbelieve in white swans. With non-human actors there is more room for white-swan-ism, and nothing I've said means that a consensus should be ignored simply *because* there is a consensus. There is genuine expertise in many fields, as well as a herd culture in many expert cohorts. Each issue should be assessed on a case-by-case basis. Anyone who denies the reality of anthropogenic climate change (the case for which is ratified by an overwhelming consensus) is misguided. As is anyone who assumes that the neo-Darwinian paradigm is a completely adequate explanation for everything in earth and heaven.

Bell curves assume that everything can be measured. Nothing significant about humans can be measured. I can't measure my love for my children, or the significance for me of the Bach cello suites, or the value of the life of the beggar I've just ignored.

Centres assume that there are fixed points from which measurements can be taken. There are no such points. There are points, as we've seen, from which we get better views than others. Multiply the viewpoints and you get a better picture. But there is no one viewpoint that gives a perfect view. And if there were, it would be infinitely far from the centre of the thing viewed. We shouldn't try to reduce things that can't be reduced. We can't reduce humans to chemicals, statistics, or anything else. In fact we can't reduce anything to anything.[8] Everything is only itself.

Statistics don't work for humans because we are, at root, unpredictable, imponderable creatures. We are daily making ourselves more ponderable, mind you. Adidas sweatshirts don't bode well for our continued imponderability. But at the moment even the Adidas sweatshirt phenomenon is fantastical – the sort of conundrum that can be explored by novelists, but not economists.

Much of this book has looked at some of our cryptic characteristics. It is worth contrasting the statistician's and economist's view of humans with that of the acknowledged experts in individual mystery. The statistician's and economist's view, as we've seen, is drawn from their analysis of cohorts, and is based on the elementary category error that an individual in a cohort of 100 has the weight of the total cohort divided by 100.

The great observers of humans do not fall into that error.

Those great observers include Jonny Foster. When he was five, he invited a friend for a sleepover. I hadn't met the friend before.

'What's Simon like?' I asked.

It was a stupid question, but I hadn't realized quite how stupid.

'He's not like anything,' came the reply.

Well, of course he wasn't. Nothing is like anything else. Similes are as fatuous as metaphors are fundamental.

C. S. Lewis, in agony after his wife's death, struggling to make sense of the reality that had cruelly invaded, wrote: 'All reality is iconoclastic. The earthly beloved, even in this life, continually triumphs over your mere idea of her.'[9]

Our ideas of everything are mean and shrivelled. The most accurate and exalted fall pitifully short of whatever it is they try to convey. That is not so much an indictment of thought and language as a compliment to reality. 'All names fall short of the shining of things,' wrote Andrew Harvey.[10] That is because everything shines so dazzlingly.

If our words fall short even of the rotting lemon at my feet in the courtyard where I'm writing this, let alone of the spider that is bundling up a wasp in front of my laptop, how much further do they fall short of the olive-harvesters who are having their lunch break and singing songs and drinking wine in the grove over the wall?

Though words fail, they can – if they're used well – do a far better job than algorithms. When they are used well, they are glorious. When they are used well, whether of humans or mice or grains of sand, they always express the contingencies at the heart of the cosmos and of human experience of the cosmos, and which are the prerequisite of wonder, and so of reverence.

'All philosophy begins with wonder,' said Plato – a dictum revised by modern analytic philosophy to read: 'If you take things apart sufficiently you'll find there's nothing at all to wonder at.' If you take things apart without putting them back together, says the right hemisphere, there's something radically wrong with your head, and you'll misunderstand the world radically. Understand it, and you'll know it can't be understood. That the Tao that can be spoken is not the eternal Tao; that if you think you understand quantum mechanics you've not understood quantum mechanics;[11] that the world is in a grain of sand, that you can hold infinity in the palm of a hand, and know eternity in an hour.[12]

Leonardo took things apart more thoroughly than anyone in history. His understanding of the role of turbulence in the operation of the heart valves, derived from days watching rivers and nights chopping up cadavers, is a masterpiece of both fluid mechanics and functional anatomy. He is bigger and more lasting than professors of fluid mechanics and anatomy because he put things back together again, multiplying and deepening the wonder: the wonder of wholeness rather than of fragments.

The business of the arts is to deal with uncertainties, because it is their business to deal with reality, and all reality is uncertain in a way that cannot be captured by a bell curve. '[A]t once it struck me what quality went to form a Man of achievement, especially in Literature, and which Shakespeare possessed so enormously,' wrote John

Keats to George and Thomas Keats. 'I mean Negative Capability, that is, when a man is capable of being in uncertainties, mysteries, doubts, without any irritable reaching after fact and reason . . .'[13]

Why is Shakespeare incontestably great? Because his characters and he himself confound all categories, conform to no theories, are unpindownable, and don't occupy or state any central position. His kings are not consistently kingly, his murderers often tearful and scrupulous, his roisterers reflective, his despots wounded children, his tragedies comedies and his comedies tragedies.[14] In short because he tells it like it is, and tells us how we are. His epistemology and anthropology are better than those found in philosophy and anthropology departments.

This, as well as the experience of, well, *experiencing* – of walking round and talking to people – makes the projects of modernism and postmodernism seem old hat and tiresome. Beckett makes us watch interminable plays about humans, which are all about how inadequate plays about humans are. Yes, I want to shout, long before the end of Act 1 Scene 1, but anyone who has met Falstaff or been a child or spoken to a drunk knows all that. Do I really have to stay?

Blake loathed the Newtonian picture of a predictable, clockwork universe. The reasons for this hatred are much debated: were they moral? Aesthetic? Theological? Scientific? He would have hated this debate. He would have denied its legitimacy, partly because those categories are non-sensical, or at least overlapping. How can something immoral be beautiful? How can something created by the creator of science be unscientific? But his objections coalesce into one: Newton's picture was detestable because the cosmos and its creatures are just *not like that*. They are contingent and unpredictable. Contingent because everything is related to everything else and dependent on it; unpredictable partly because of that infinite web of contingency, and partly because of the fact of genuine freedom.

There are no fixed points. In trying to insist that there are, the centre is pleading desperately, and in the teeth of the evidence, for its own existence. Centres don't exist, so let's not pretend that we can live there. Average humans don't exist, so let's not pretend that we are one, or that we're married to one. The only places we can live are right over the edges of the bell curve, in the real, edgy world made partly of matter (whatever that is) and wholly of enigma.

What, then, is the centre? There's no such place. Or if there is, no one can live there. Yet this fiction on a fraudster's tongue is infused with malevolent personality, and though it doesn't exist, its armies and civil servants certainly do. Vilify it as I have, and the centre will try to get you.

The strategies the centre uses to quell dissent tell us a lot about it. We'll look at them next. You're safer if you know your enemy.

The War on the Edges

In Portugal they will have to prepare papers to say who they are. How can anyone do that? How can anyone say who he is?

V. S. NAIPAUL,
Half a Life[1]

I MET DAN ON A mountain in India. He was sitting on a rock, pulling leeches off himself. There was one on his back he couldn't reach. I helped him, he gave me a swig of very good whisky from a hip flask that had belonged to his foxhunting grandfather, and we spent the next fortnight walking, talking, laughing and eventually weeping.

He came from Croydon, was translating the Upanishads into Gaelic, meditated for two hours before dawn, could improvise for hours in dactylic hexameters, had swum the Hellespont twice, had spent four foodless nights with a broken ankle in the Bolivian jungle reading *Winnie-the-Pooh*, was planning to compile a book of Inuit seal recipes, had memorized a good half of Scott's Border ballads, had a girlfriend who would have made Aphrodite look like a tortoise, and earned his living by preparing and mounting animal skeletons.

He lived in a bothy in Ardnamurchan, a stone's throw from the sea. I visited him there. We kayaked together up the coast, trawling for mackerel, swam with the puffins, stared into the driftwood fire, crawled after the red deer, and wondered where he should publish his PhD thesis – which was on syphilis in fourteenth-century Bohemia.

Once, very drunk, he'd applied for a research fellowship, and sent off a highly sanitized application (with a fraction of his publications and none of his real achievements) before he'd sobered up. To his surprise and amusement, he'd been shortlisted, and since he had to be in London anyway to help a friend pulp apples for cider-making, he borrowed a suit and went along for the interview. The questions, he said, were so ludicrous he'd started to laugh, and couldn't stop, and he staggered out of the room still laughing, and was still laughing on the train.

Out of pure scientific interest (he said), he'd sent off unsanitized applications to act as controls. They included all his publications and an honest account of his life. He never got a reply.

Without exception, the most interesting and downright talented people I have known have repudiated the centre and been repudiated by it. The interest and accomplishment of people are in direct proportion to their distance from the centre. No, that's not right: they vary by the cube of the distance from the centre.

At some level the centre knows this. There is no other way to account for the venom of its attack on the edges. The centre knows that it is intellectually, aesthetically and spiritually inferior, and can't bear the thought. Wipe out the opposition, it thinks, and its own inadequacy won't be so humiliatingly obvious.

There's a war on.

The centre seeks to remake the world in its own image – the image of those infernal siblings: the metropolis, the monopoly, the monoculture and the pint of homogenized milk.

The centre seeks to extend its franchise. The words for that, in this context, come from cancer biology and the dairy industry: the centre seeks to metastasize and homogenize. If everything and everyone is the same (whatever that sameness is – it might be possession of the same brand of sweatshirt, addiction to the same brand of music, or whatever) it will be far easier for the centre to metastasize. It's terribly easy to substitute a universal passion for a phone for a universal love of a particular centre or a particular Self. Uniform people are manipulable people. So the centre loves geographical hubs which abolish or threaten the good type of edges between people. It peddles the lie that homogeneity is liberal and compassionate – whereas if everyone is the same, the love of others is a type of self-love. It shudders as it remembers De Gaulle's declaration that a nation with 365 types of cheese cannot be governed.

The centre loves it when the city seeps out into nasty, ungovernable country, putting identical houses on the ruins of old forests. It hates trees, come to that, for they live longer than we do, and so challenge our hubris, and don't come in neat shapes. It hates the wild, which won't come quietly – which is another reason why the centres are almost always *urban*, and usually located in national or regional capitals. Howard Jacobson, in *J*, describes how a fictional regime very like our own outlaws jazz, for if you start improvising in one sphere, where is it going to end?[2] The wild improvises all the time.

The centre loves its abstractions, its categories and its bureaucracies. It is pitilessly technocratic (even when it is notionally artistic), for technology can be channelled, taxed and controlled as the wild, edge-crossing humanities cannot. The centre seeks to convince us that its own algorithmic pastiche of reason is superior to the way real, edgy, enfleshed people reason. The algorithmic view is all about calculation. Real reasonableness is not: think of the 'reasonable man' test in the law – which turns on what jurors with experience of buses, tired children, testy spouses, messy sex and plum pudding make of a situation. The algorithmic view, as we've discussed, rests on the preposterous presumption that fundamentally incalculable things – such as human desires, fears and intentions – are calculable. Because only calculable things are controllable, the centre has to pretend that everything is calculable – or make it so.

The centre fears and distrusts emotions themselves, and seeks to excise them from its decision-making. We all know – for the moment, though it won't be long before it is forgotten – that this cannot be done in relation to any tricky human decision-making. Emotions, writes Martha Nussbaum, are 'suffused with intelligence and discernment' and 'contain in themselves an awareness of value or importance.' They are not 'just the fuel that powers the psychological mechanism of a reasoning creature, they are parts, highly complex and messy parts, of this creature's reasoning itself, and so cannot 'easily be sidelined in accounts of ethical judgment, as so often they have been in the history of philosophy'[3] – a sidelining done always by the centre, and particularly since the Enlightenment.

Emotions, complexity and the very substance of humans themselves are all the business of the humanities. The centre recoils from the humanities as if from a rabid dog. It defunds them if it can, and if that fails, denigrates and neuters them.

One ploy is to get the humanities to see themselves as 'researching', in precisely the same sense as one researches in science or medicine or engineering. Thus there is a pathetic thirst in English departments for measurable outcomes. Am I 'researching' if I sit in a library and write about George Eliot? I must pretend so. Since scientific-type innovation in the humanities simply isn't possible, people increasingly don't bother to pretend. The result is extreme conservatism: good scholarship is adding footnotes to footnotes. Great scholarship is adding footnotes to footnotes to footnotes. The scholarship is palpably boring and pointless, and so the technophiles smile, rake in the research grants for the next generation of chatbots, and close down the English departments, and that's just the way the centre likes it.[4]

The centre has no moral scruples about the use of violence in its dealing with the troublesome edges, but since violence may have incalculable consequences (producing unpredictable, edgy, radiating ripples), and hence may be a less efficient means of control, the centre tends to use violence as a last resort.

Usually it tries first to convince the edges of the error of their ways. This has had limited success, particularly since edge-people remain people. Even if they don't read John Donne or Nabokov, they still have dying parents and malfunctioning bowels. Only in rarified bubbles in Silicon Valley and academia, where cognitively hypertrophied people, out of vanity, convince themselves of their bodilessness and their immortality, is it possible to believe that humans are secure; that they aren't always walking a tightrope over a yawning chasm.

Much more effective is the second strategy: assimilation. This doesn't depend on effecting any change in mind; just on convincing the subject that resistance is hopeless, or just too exhausting, and that they might as well come to the centre, where they'll be made to feel welcome and plied with goodies. The centre flatters to allure, and however false we know the flattery to be, and however uncongenial we know the centre to be, we are still drawn moth-like to the flame. In C. S. Lewis's *That Hideous Strength*, the best modern account of the centre's strategy of assimilation, the young, ambitious, gullible academic, Mark Studdock, is drawn almost fatally towards the epicentre of a diabolical conspiracy. In a rare moment of insight he recalls how, as a child, he craved the company of the Inner Circle, despite knowing that its delights were illusory. He remembered how he had tried to convince himself that he enjoyed the company of the athletic demi-gods of the school. He remembered 'laboriously reading rubbishy grown-up novels and drinking beer when he really enjoyed John Buchan and stone ginger', the pathetic efforts he made to learn the language of each new circle that attracted him, and 'the almost heroic sacrifice of nearly every person and thing he actually enjoyed'. All this 'came over him with a kind of heart-break. When had he ever done what he wanted? Mixed with the people whom he liked? Or even eaten and drunk what took his fancy? The concentrated insipidity of it all filled him with self-pity'.[5]

Don't be fooled. Laugh along with Blake at the conceit of the centre (for whom 'Fortune' stood proxy). 'I laugh at fortune', he wrote. 'The Goddess Fortune is the devil's servant, ready to kiss anyone's arse'.[6]

Often the centre pretends that the move is not particularly significant, or is temporary. At the very start of the Genesis story, the disobedience that sought to blur the divinely ordained divisions between things looked undramatic at first. There was no hail of fiery darts; no kicking of newly created backsides with demonic hoofs.

Just twilight – an almost imperceptible seeping of darkness into light – instead of a clear demarcation of the two; and insubordination so subtle that it is visible only in Hebrew. Only light did what it was told: 'Let light be,' commanded God. 'Light be,' came back the report. From then on, there was mounting insubordination. At first it was slight. 'Grass grass,' God told the earth. But that's not *quite* what happened: the earth, instead, with just a touch of rebellion, 'put forth' grass. The created order then seemed to gain confidence from its cockiness. By the time God asked the waters to 'bring forth' the water animals, the earth simply refused. (The sea and God are often at odds.) God was forced, for the first time, to create directly: 'So God created the great sea monsters and every living creature that moves, of every kind, with which the waters swarm.'[7]

Often the centre convinces the edge-people that they can hang on to their edginess, but that in the interests of that edginess, they should use the methods and power of the centre. It is always a lie. The centre plays on fears and insecurities, hyping them, and implying that there is safety only within the thick walls of the metropolis.

War is the centre's best friend. Every autocrat knows that if things are bad at home, it's shrewd to start a war abroad. It consolidates the power of the centre. It substitutes national themes for personal ones. It makes people wave flags rather than talk or make love or read books. Knowing this, Lawrence Durrell, in the Second World War, plotted a magazine about 'those personal landscapes' which obstinately continue to exist (despite all the flag-waving) outside national and political frontiers. On the edges, that is, where all reality is. What troubled him most about the war was the 'canalization of thought and response'[8] – the shepherding of real people and ideas into the containers provided by the centre for the purposes of control.

If assimilation fails, the centre cold-shoulders and denigrates.

It characterizes edgy states of mind as pathological. It declares that eccentrics are diseased. The disease need not be manifested in shamanism or clairvoyance or enthusiasm for mindfulness practice. Simple disagreement with an orthodox view is a symptom sufficient to justify the diagnosis of ideological leprosy. This is old-fashioned heresy, of a kind familiar to religious inquisitions and totalitarian projects. Dissent from the received wisdom may or may not entail a trip to a literal gulag for re-education, but it will certainly mean a journey away from the comforting hearth where the obedient warm themselves. It will mean loss of academic tenure or other preferment. Since academic publication depends on the iniquitous process of peer review, only those whose views accord with the orthodoxy will be read. It makes academia suffocatingly conservative. Institutions purporting to be built of fearless Enlightenment scepticism, concerned solely with the truth, are concerned only with perpetuating a view of the truth that accords with the received canons. It is far more religious than the Church, against whose fossilized ideology the Enlightenment was originally a reaction. The sound of the academy should be the scream of old paradigms being interrogated and killed. Too often it's the sound of meekly recited creeds, hymns to hoary orthodoxies, sycophantic psalms to the saints of the discipline, and calls for the martyrdom of infidels.

Recall the lynching of the biologist Rupert Sheldrake, who had the Enlightenment-inspired audacity to suggest, in his 1981 book *A New Science of Life*, that there might be a new way of looking at issues of causation in the natural world. Sir John Maddox, then editor of the canonical journal *Nature*, was so outraged that only the language of the Inquisition would suffice. The book was, he fulminated in an editorial, 'an infuriating tract' and 'the best candidate for burning there has been for many years'.[9] This wasn't just a rush of blood to the head: interviewed years later about his outburst he declared that he had been 'offended' by the book, and that Sheldrake's work could be 'condemned in exactly the language that the Pope used to condemn Galileo, and for the same reasons. It is heresy'.[10]

Heresy against what?

The Cambridge physicist Brian Josephson was awarded the 1973

Nobel Prize for Physics for his description of the 'Josephson Effect', a phenomenon in quantum tunnelling which, amongst other things, involves the flow of current across the junction of two superconductors even when there is no voltage drop across the junction. Yet when Josephson turned to consider whether effects such as telekinesis and telepathy might be real, the Establishment turned against him. He was shouted down at conferences and his articles were refused by journals. There were things that simply could not be thought: let him who thinks them be anathema. Free-thinking about thought and modes of consciousness is particularly anathematized, for that goes to the root of centrism: the centre presumes a certain way of perceiving the world, and decrees that that is the only way.

Rupert Sheldrake himself, denied a place in one of the Establishment's institutions, conducts his extraordinary research in a laboratory set up in a bedroom of his London house, often assisted by his son, the mycologist Merlin Sheldrake. He stands in a grand tradition of independent researchers – all of whom would be ignored by contemporary academia on the grounds that they did not have appropriate headed notepaper. Think of the Augustinian friar Gregor Mendel, whose experiments with pea plants laid the foundations of genetics. Think of Gilbert White, the gentle parson-naturalist of Selborne, who (because he ranged across many disciplines, fascinated by everything from earthworms to swift migration) became the father of ecology – the science of biological entanglement. Think of Darwin himself, an avid reader of White, pacing round the garden at Downe House, checking his wormeries, turning over skulls, chipping away at rocks, stroking the skins of the birds he had brought back from the adventure of the *Beagle*. Think of James Lovelock, who, having worked on the important question of whether hamsters could be deep-frozen and then revived, formulated his Gaia hypothesis, which saw the Earth as one organism.

None of them would be given the time of day by any of the pontifical figures of modern biology. Science has become a profession rather than a quest for the truth. And professions *control* their members. There are things that professionals cannot do or say. If you want a place in the centre, you can't be seen to be flirting with anyone at the edges.

If the cold shoulder fails, the next stage is the raised hand. Sometimes it is used to force compliance; to drive the errant back to the centre. We've seen that the demons, alarmed and outraged by the damage St Anthony was doing to their legions when he clashed with them in the desert, tried to drive him back to the metropolis, knowing that much of his edge-power would be neutralized there. Governments of all eras try to herd edge-people into manageable cages. In the thirteenth century BCE, Egypt energetically tried to subdue the shifting populations of Shashu and Apiru. In twenty-first-century Britain, to be without an address is to be invisible to governments, election officials and banks. It's as if institutions see humans as made of bricks and mortar. No bricks and mortar, no existence. Government funding in all modern jurisdictions is typically directed towards places where there are lots of bricks and mortar: to the cities. Places beyond the city boundary are often forgotten. Deliberately. If you want your slice of the fiscal cake, move back to the metropolis and be controlled.

And if the errant won't go? Well, then there are no limits.

John Pilger relates a story told by an old woman in Victoria. Australian Aboriginal babies were buried alive, with only their heads above the ground. The brave British invaders then had a competition to see who could kick the babies' heads off furthest. The rest of the day was spent raping the women and torturing them to death – typically by sticking spears into their vaginas until they died. The men's hands were tied behind their backs. Their penises and testicles were then hacked off. The Brits watched as they ran around screaming until they bled to death.[11]

There was no sin in this, for Aborigines were not considered fully human. 'I do not think ... that the doctrine of the equality of man was really ever intended to include racial equality', proclaimed

Sir Edmund Barton, Australia's first prime minister. 'There is that basic inequality. These races are, in comparison with white, unequal and inferior. The doctrine of the equality of man was never intended to apply to the equality of the Englishman and the Chinaman ... Nothing in this world can put these two races upon an equality. Nothing we can do by cultivation, by refinement, or by anything else will make some races equal to others.'[12] The British had declared that Australia was *Terra Nullius*: nobody's land, for there were no true humans there. Declarations from the centre were definitive. The Aboriginal peoples were on the far side of every single edge visible from the gentlemen's clubs of St James's. In particular they walked long distances in bare feet and valued modes of consciousness other than those on show in a *Times* leader column.

Wandering itself – the continual crossing of frontiers – has disconcerted sedentary centrists since the days of Cain and Abel, often with the same result. Blood on the ground. The Nazis killed between 250,000 and 500,000 Roma. Siberian shamans – those incessant commuters between dimensions – were mocked by the Soviets for saying they could fly, and thrown out of helicopters to give them the chance to prove it. If your address is 'A tent, wherever I happen to end up when it goes dark', rather than '10A Acacia Close', you're likely to be disenfranchised. If you're in the margins you're *legally* marginalized.

Of course edge-ideas have long been dangerous, and what is edgy may depend on when and where you are. To believe that the Eucharistic bread and wine were the body and blood of Jesus was sometimes as deadly as believing they were not. Trofim Lysenko, director of the Institute of Genetics at the Soviet Academy of Sciences, was a hysterical advocate – for political reasons – of Lamarck's theory of the inheritance of acquired characteristics. He sent to a firing squad or the gulag many who believed that Mendelian genetics was right.

'Thou shalt not suffer a witch to live', screams the book of Exodus,[13] a scream often grafted into statute. Here, for instance, is part of an 'Act against Conjuration, Witchcraft and dealing with evil and wicked Spirits' (1604):

> . . . if any person or persons, after the said Feast of St. Michael the Archangell next coming, shall use, practise, or exercise any invocation or conjuration of any evil and wicked spirit: or shall consult, covenant with, entertaine, imploy, feed, or reward any evil and wicked spirit, to or for any intent or purpose; or take up any dead man, woman, or child, out of his, her, or their grave, or any other place where the dead body resteth; or the skin, bone, or any other part of any dead person, to be imployed, or used in any manner of Witchcraft, Sorcery, Charme, or Inchantment, or shall use, practise, or exercise, any Witchcraft, Incantment, Charme or Sorcery, whereby any person shall be Killed, Destroyed, Wasted, Consumed, Pined, or Lamed, in His or Her body, or any part therof; that then every such Offender, or Offenders, their Ayders, Abettors, and Counsellors, being of the said offences duly and lawfully Convicted and Attainted, shall suffer paines of death as a Felon or Felons, and shall lose the priviledge and benefit of Clergy and Sanctuary.[14]

In Europe and North America, from 1400 to 1775, between 40,000 and 60,000 supposed witches – 80 per cent of them women, and many of them old – were executed.[15] Mrs Bogg wouldn't have had a chance. You have to feel very insecure to murder on that scale.

Simply killing an edge-person may not be enough. The offence might need to be marked by eternal death or punishment. There was a systematic attempt to erase Akhenaten's name from history. The erasure was often literal. Tomb 188 in the Theban necropolis is the tomb of one of Akhenaten's high-ranking servants. Akhenaten's image has been obliterated – a violation designed to affect Akhenaten's enjoyment of the afterlife. It could be worse: excommunication in many Christian traditions is designed to hurl the apostate into an eternally blazing lake of fire.

We have seen many examples in this book of the centre privileging the states of consciousness with which it itself is most comfortable, and which are most predictable and easiest to control. These are also the states least responsible for making us what we are as a species, and as individuals. We have seen many examples of heterodox consciousness being met with ostracism or the death penalty.

There is one final strategy: that of saying that non-orthodox consciousness is not consciousness at all and that accordingly (since orthodox consciousness is the lodestone of true agency), decisions must be made on behalf of the non-capacitous person (who is typically dubbed a 'patient' to make it all the more obvious that non-orthodox consciousness is pathological).

Take an extreme example, which perhaps shows what is really going on in less dramatic cases: a patient in Permanent Vegetative State (PVS) – a profound and perpetual coma in which, goes the orthodoxy, no sensation, painful or pleasurable, will ever be possible. The patient will breathe spontaneously, but will need to be fed – usually by a nasogastric tube. Note the language. The 'patient' is tantamount to a vegetable. The life (if any) of a vegetable can be unproblematically ended. We don't think twice before boiling and eating a courgette. (A hunter–gatherer would demand prayers, apology and reparation. For us, though, the courgette is morally irrelevant.)

We have no idea whatever about the internal life of a patient in PVS, except that it is very different from our own. It is solely this difference that is used to justify a common fate of PVS patients – death by withdrawing their food and fluids. A human who does not have consciousness like ours, in other words, is better off dead. The lawyers put it less coarsely: the patient has no interest in continued biological survival.

But what's so great about our type of consciousness?[16] Everything we know from dreams, psychoanalysis, our non-cognitive responses to music, our subterranean intuitions and so on suggests that most of our real living isn't done in that tiny suburb of consciousness whose existence alone, in medical and legal orthodoxy, justifies continued life. For all we know, the damage – whether by stroke or car accident or whatever – that has caused the PVS might have slackened off Huxley's 'reducing valve', allowing into the brain a great surge of information and sensation. The patient, uninhibited for the first time by the jealous and aggressive policing of workaday consciousness, might be having the time of their life; might be thinking pityingly of us, and of their own past life, as sadly straitened.

Amanda Feilding, one of the pioneers of psychedelic research, has posited that psychedelics might activate more of our brain than we usually use, and might enable communication between parts of the brain that normally function more or less autonomously, in silos, behind their boundary walls. Psychedelic use, on this model, would entail transcending boundaries. Something similar might happen in profound disorders of consciousness such as PVS or minimally conscious states. If that's right, the 'disorder' is a pathology only of a usually pathologized brain. The pathology of a pathology is what we normally call healing. The point of cancer treatment is to kill aberrant cells. The point of self-realization might be to knock down pathological partitions between our compartments so that we can occupy all of ourselves.

The archaeologist Steven Mithen characterizes Neanderthal neuropsychology as 'rigidly compartmentalized' – starkly distinct from the minds of Upper Palaeolithic *Homo sapiens*. We're becoming more Neanderthal by the day.[17]

The centre guards itself very diligently. It sees its main threats as coming from the edge-people. In this at least it is quite right. This is because all its inhabitants, being human, are at root edge-people. The poor centre: every single one of its citizens has an innate tendency to insurgency.

The edges, too, have a robust solidarity. I suppose it's not surprising since everything is fashioned from them. The nomad helps the nomad; the dying support the dying; the destitute give all to the destitute (and find that they themselves become rich). But there are other, stranger forms of solidarity too.

Many years ago, after that bruising, chafing trip in the Sudanese desert – in which, by my own presumption and arrogance, I'd lost a treasured friendship, and very nearly lost my life – I flew back to London at night from Khartoum. I didn't want to come back. I've never been one for the fleshpots, and though spitting camels and torrential diarrhoea had drained the desert of its immediate appeal, I'd far rather have been back there than in the London drizzle.

Though I have long legs, I always ask for a window seat, and after the deep dark of the Mediterranean I saw the first sparse lights of Europe. I saw what were obviously very high mountains, spotted with tiny villages, miles apart, and a thin ribbon of light where the land slid into the sea.

Now that, I thought, is a place I could live. Those villages will each have a tiny taverna which will only give you what happens to be in the pot, and one colour of home-made wine.

It was the very south of mainland Greece. At the top of one mountain was a completely distinctive and unforgettable pattern of lights. Unforgettable, but I forgot about it for years.

Many years later my mother died – with a stoicism that appalled me. She slid into the dark like those Greek mountains into the sea.

A fortnight after we'd burned her, a friend took me off to Greece to try to make things better. I had no idea where we were going, and didn't care. The friend had fixed it all. I was numb, but as he drove me (the worst company in the world) through the spring, sensation began to return. First there were pins and needles; then good, honest mental pain; and then I started to scream underwater as we swam through thickets of sea grass, and then I could laugh laconically at dinner and finally sob silently in my room.

The spring flowers, in that crystalline light which shows everything, including yourself, the way it really is, were like the Platonic forms of flowers. They gave no assurance of resurrection. If anything they taunted me, the bastards, because they were alive and she wasn't, and she mattered more than any fucking flower, and her only public acknowledgement that she was going to die was to say, in a matter-of-fact way, as you'd say that the bin men would be round later: 'I won't see the daffodils next year.' There was no reassurance of resurrection in the fact that the flower seeds had sort of died, and yet here the flowers were. I'm smart enough to know the difference between a metaphor and my mum.

Anyway, I looked out of the window over the sea and into a big ravine. On the lip of the ravine hung a village, most of it just round the corner, out of sight. I couldn't be happy anywhere now, I thought, for the world is trashed. But I'd be least unhappy there.

Many years after that I bought a house in the village where I'm writing this. I had no idea, until two years after I'd bought it, that this was the village I'd seen just after my mother's death. I'd thought we'd stayed over the other side of the peninsula. But that was nothing compared to what happened a year later.

One dark winter afternoon, walking in the mountains above the house, I came upon that unforgettable configuration of lights I'd seen from the Khartoum plane. I walked around it five times to make sure. There was no doubt.

All this is very weird and very wonderful. I like to think it shows that the edges look after their people.

The centre, ultimately, has no chance against *us*, however many of us it dispossesses and kills, for the whole cosmos is made of edges, and you can't win against the cosmos.

Epilogue

First Man, Edge Man, Everyman

AT THE END OF this rambling journey there is an almost irresistible temptation to produce an executive summary: to tie up loose ends.

I am going to resist it. If loose ends are left dangling, there are more edges on display. I've insisted that that's a good thing, and I shouldn't recant now. Instead I will tell you what I did last night.

I drove again to the dirt road above the First Man's (well, the First European's) cave, and walked down under the eyes of the limestone fortress, of the hunting swallows, and the goats. The cicadas rasped, and out at sea Giorgos was hauling in red snapper. I had a bottle of local brown wine in my backpack, and two glasses.

I brushed a scorpion off my usual rock on the clifftop and sat down. I took out the bottle and the glasses and the sheaf of articles about the First Man's skull and put them beside me. I'd been used to reading those articles here, as if the Man himself might tap me on the shoulder and question the Carbon-14 results. But tonight I didn't read them. There had been too much reading.

I poured out two glasses of wine: one for me and one for him. I clinked them, said, '*Yamas!*' – 'For us!' – though I knew he didn't speak Greek, and drank first my glass and then his, thanking him for it.

If a swallow generation is three years, there are 70,000 generations between the swallows he saw looping over the cliffs and those I saw. If a human generation is twenty years, there are 10,500 generations of us between him and me. This is not so many. The sun he saw sinking over the edge of the world burned slightly more brightly than the one I watched that night, but not so you'd notice. He knew, just as I know, that there are only so many times that the sun can fall and climb before we go over the horizon to meet it, like Reepicheep. I raised my glass and his to the sun. There was not much in the bottle now. I wished I'd brought another. We were getting through it well, the two of us. It was made of a very old variety of grapes, Agiorgitiko – St George's grapes – and it has helped to kill, or see off for the night, many types of dragon.

We had drunk too much for me to drive home, and it would have been rude to go anyway. It was a clear night, but I had an old tweed jacket and a Kenyan *kikoi* that makes the Greeks laugh at me, and the red soil would give out heat like a woman until morning.

This was the time, at the seam of the day and night, when things happened, when moods changed, when looking straight at anything didn't work, but the fringes of vision were busy and forgotten noses and ears and senses without names began to remind me they were there; that they had always been there; had always been contributing. An owl bounced along the seam, as if it wanted to stitch day and night together. It was a Little Owl – *Athene noctua*, the bird of the goddess Athena, the goddess of wisdom, for wisdom is at the meeting of the edges. Its eyes are proportionately much bigger than mine, and its ears much better. It gathered a lot more information about this clifftop than I did. It dropped, scrabbled in the oregano, and rose with a struggling mouse as the last edge of the sun vanished. But there was still light from over the edge of the world; quite enough to pour the last of the wine into the glasses.

We had come a long way together, the First Man and I. I am very grateful to him. He has taught me much. I used to think he was unusual. But in the journey described in this book he has been joined by Shakespeare, Leonardo, Akhenaten, giggling vicars, artistic toddlers, tortured saints, shamans with antlers sprouting from their brows, Upanishad-translating skeleton-mounters, dead parents, prime ministers, wart-faced healers, innovative bacteria, heretical biologists, insecure dons, injured seabirds, polyglot monks, dozing meditators, frustrated demons, psychopaths, wannabe bankers with

240

hoofs, comatose patients, sacrificial Bedouin, big-wave surfers, fulminating fundamentalists, Egyptian street kids, vampire finches, gossiping electrons, and hordes of unruly ideas. And in fact by everyone and everything.

The First Man is not unusual at all. He is all of us, and our thriving depends on us knowing it.

I chinked the glasses again, put them down, curled up on a bed of sage and slept until I was kicked awake by a shepherd. The sun was back.

I threw the glasses into the sea, bowed to the First Man, and ran up the hill to send off this book to my patient editor in London.

Acknowledgements

Three people have been particularly influential in shaping the thoughts in this book.

The first is Iain McGilchrist. He is one of my best friends, and many of my happiest and most formative times have been at his house on Skye. My personal and intellectual debts to him are immense. The intellectual debt, at least, is obvious in this book. But there are some important dissonances between his song and mine. To those who know his work, those dissonances will be clear. To those who do not, I need not trouble you.

The second is Andrew Graham-Dixon, the most penetrating art historian alive. He was generous with his time and wisdom, and in talking with him and reading his work, many pieces in the strange puzzle of the edges slotted into place.

The third is C. S. Lewis (whom of course I never knew), who is quoted here much more than any other writer. I had no idea until I started to write this book that so much of his work was, in one way or another, about edges. I shouldn't have been surprised: an Ulsterman living for most of his life in Oxford couldn't help seeing the universe from an edge, and an Ulsterman who, it seems, had read everything, and remembered it all, could hardly help noticing that this perspective was shared by so many others that it seemed to be *the* human location. And an Ulsterman who could say in three words what the rest of us take a book to spell out could hardly be left uncited.

I owe debts to many others. Some are mentioned below, many are not – including massively influential people I met on the road and whose names I never knew.

I am hugely grateful to the following:

Jan Abbink, David Abram, Alexia Adams, Theo Bargiotas, Joshua

Barley, the late R. J. (Sam) Berry, Magnus Boyd, Rabbi Eli Brackman, Dominic Burbidge, John Butler, Rachel Campbell-Johnston, Nuno Castel-Branco, Laura Carew, Peter Carew, Sonya Clegg, John Cooper, Margaret Cooper, Vicki Cummings, Barry Cunliffe, Colette Dewhurst, Apostolos Doxiadis, Melina Dritsaki, Robin Dunbar, Marguerite Dupre, Brian Earp, Steve Ely, Avi Faust, Israel Finkelstein, Kate Foster, Clive Gamble, Yossi Garfinkel, Jan-Bart Gewald, Shimon Gibson, Federica Gigante, Petra de Goeij, the late David Graeber, Father Iain Graham, Jay Griffiths, Jo Hamilton, Jonathan Herring, Sally Hope, Tony Hope, Barry Howard, Gill Howard, Nelleke IJssenagger, Jeri Johnson, the late Metropolitan Kallistos, Pat Kaufman, Max Kestner, Paul Kingsnorth, Sarah Knowles, Marinos Kyriakopoulos, Klara Leppo, Andy Letcher, Abigail Lloyd, Michael Lloyd, Andy McGee, Nigel McGilchrist, Steven Mithen, George Monbiot, Mariam Motamedi Fraser, James Mumford, the late Jenni Nuttall, Claudia Oik, Stamatoula Panagakou, Michael Parker, Vangelis Pastelakos, Metaxia Pavlakou, Paul Pettitt, Theunis Piersma, Costa Pilavachi, Andrew Pinsent, Kristen Poole, Keith Powell, Jonathan Price, Jill Purce, the late Steve Rayner, Glyn Redworth, Caroline Anjali Ritchie, Andrew Roe, Julian Savulescu, Rick Schulting, the late James C. Scott, Emma Searle, Berny Sebe, Martin Shaw, Merlin Sheldrake, Rupert Sheldrake, Claire Smith, Mike Smith, Katerina Stathatou, Andrew Steane, Barney Taylor, Adrian Thirkell, Julian Thomas, Peter Thonemann, Caro Thouless, Chris Thouless, Ruth Thrush, Sir Rick Trainor, Johan van de Gronden, Mark Vernon, Martin Voss, Hinny Wass, Joe Wass, Emily Watt, Jimmy Watt, Melanie Watt, Harry Wels, Mark West, Sue West, Father George Westhaver, Dominic Wilkinson, Matt Wolf, Ahmed Yousef, and any Fellows of Exeter College, Oxford not already mentioned.

John Stathatos kindly gave permission to reproduce his splendid translation of *The Satrapy* (p. 58), and his photographs of walls on Kythera expound my thesis so well that my words are probably unnecessary.

My brilliant editor, Alex Christofi, believed in this book when there was no reason at all to do so. His kindness and patience made the book a pleasure to write, and his acute editorial eye has made this a much less bad book than it was.

My splendid copy-editor, Becky Wright, ruthlessly culled the herds of infelicities I'd allowed to graze throughout the text.

As ever, I salute my agent and friend, Jessica Woollard: a constant support and a wise counsellor.

My children Lizzie, Sally, Tom, Jamie, Rachel and Jonny are my great teachers. Thank you for bearing with me throughout the physical and mental absences, and perhaps even more throughout my presences, and for sometimes even talking to me though I am embarrassing.

And as for my wife, Mary: there are no words. It's all her, really. Thank you so much for it all.

There is no bibliography. I am suspicious of bibliographies, because *everything* everyone has ever read appears in some form in everything they say and write. The more obvious sources are in the end notes. I commend, though, Maria Popova's website *The Marginalian* – a superb collection of resources on edginess: see https://www.themarginalian.org.

All the characters depicted are real people or based on real people. I have changed some names and locations.

References

1   G. K. Chesterton, *Autobiography* (Hutchinson, 1936).
2   C. S. Lewis, *The Inner Ring*, 1944 Memorial Lecture, King's College, University of London.

A WORD BEFORE WE SET OFF

1   I am thinking, of course, primarily of Chatwin's *Songlines* (Franklin Press, 1987).

I: BEGINNINGS

1   Tony Judt, 'Edge People', *The New York Review of Books*, 25 March 2010. Accessed at: www.nybooks.com/articles/2010/03/25/edge-people.
2   John Betjeman, 'An Edwardian Sunday: Broomhill, Sheffield', in *High and Low* (Houghton Mifflin, 1967), p. 28.
3   For further discussion, see C. S. Lewis, *The Discarded Image: An Introduction to Medieval and Renaissance Literature* (Cambridge University Press, 1964), pp. 92–112.
4   Numbers 15:37–8 (New International Version). See too Deuteronomy 22:12.
5   For detailed discussion, see Johan Eklof, *The Darkness Manifesto: How Light Pollution Threatens the Ancient Rhythms of Life* (The Bodley Head, 2022).
6   Avraham Faust, 'The Bible, Archaeology, and the Practice of Circumcision in Israelite and Philistine Societies', *Journal of Biblical Literature* 134: 2, pp. 273–90. Cf. Max Price, 'Food and Israelite Identity', in *T&T Clark Handbook of Food in the Hebrew Bible and Ancient Israel* (Bloomsbury, 2021), pp. 423–63. For discussion of the issue of Israelite ethnogenesis, see Sara Mohr and Shane M. Thompson, eds, *Power and Identity at the Margins of the Ancient Near East* (University Press of Colorado, 2023); Avraham Faust, 'Between the Biblical Story and History: Writing an Archaeological History of Ancient Israel', in *The Ancient Israelite World* (Routledge, 2022), pp. 67–82; Mario Liverani, Niels Peter Lemche and Emanuel Pfoh, 'The "Origins" of Israel: An Unachievable Project of Ethnogenetic Research', in *Historiography, Ideology and Politics in the*

Ancient Near East and Israel (Routledge, 2021), pp. 257–73; Avraham Faust, *Israel's Ethnogenesis: Settlement, Interaction, Expansion and Resistance* (Routledge, 2016); Emanuel Pfoh, 'On Israel's Ethnogenesis and Historical Method', *Holy Land Studies* 7: 2, 2008, pp. 213–19; Israel Finkelstein and Neil Asher Silberman, *The Bible Unearthed* (Free Press, 2001).

7 See, for instance, Cato's broadsides against the Greeks. An example is in Pliny the Elder's *Natural History*, 29.7, where Cato is quoted as follows: 'Concerning those Greeks, son Marcus, I will speak to you more at length on the befitting occasion. I will show you the results of my own experience at Athens, and that, while it is a good plan to dip into their literature, it is not worthwhile to make a thorough acquaintance with it. They are a most iniquitous and intractable race, and you may take my word as the word of a prophet, when I tell you, that whenever that nation shall bestow its literature upon Rome it will mar everything; and that all the sooner, if it sends its physicians among us. They have conspired among themselves to murder all barbarians with their medicine; a profession which they exercise for lucre, in order that they may win our confidence, and dispatch us all the more easily. They are in the common habit, too, of calling us barbarians, and stigmatize us beyond all other nations, by giving us the abominable appellation of Opici. I forbid you to have anything to do with physicians.' (Trans. Karl Friedrich Theodor Mayhoff, 1906.)

8 Thucydides, *History of the Peloponnesian War*, Book 1, trans. Peter Thonemann (personal communication).

9 See, for example, Stephen G. Post, 'Altruism, Happiness, and Health: It's Good to be Good', in *An Exploration of the Health Benefits of Factors That Help Us to Thrive* (Psychology Press, 2014), pp. 66–76; Bryant P. H. Hui, Jacky C. K. Ng, Erica Berzaghi, Lauren A. Cunningham-Amos and Aleksandr Kogan, 'Rewards of Kindness? A Meta-Analysis of the Link Between Prosociality and Well-Being', *Psychological Bulletin* 146: 12, 2020, p. 1084; Nicole D. Anderson, Thecla Damianakis, Edeltraut Kröger, Laura M. Wagner, Deirdre R. Dawson, Malcolm A. Binns, Syrelle Bernstein, Eilon Caspi and Suzanne L. Cook, 'The Benefits Associated with Volunteering Among Seniors: A Critical Review and Recommendations for Future Research', *Psychological Bulletin* 140: 6, 2014, p. 1505; Netta Weinstein and Richard M. Ryan, 'When Helping Helps: Autonomous Motivation for Prosocial Behavior and its Influence on Well-Being for the Helper and Recipient', *Journal of Personality and Social Psychology* 98: 2, 2010, p. 222.

10 *Centuries of Meditations.* Traherne's dates were 1636/7–74. *Centuries of Meditations* was not published until 1908. Since then it has gone through many editions.

11 See, for instance, Larry Siedentop, *Inventing the Individual: The Origins of Western Liberalism* (Harvard University Press, 2014); Norman M. Ford, *When Did I Begin?: Conception of the Human Individual in History, Philosophy and Science* (Cambridge University Press, 1991); Eric T. Olson, *The Human Animal: Personal Identity without Psychology* (Oxford University Press, 1999); Steven Lukes, *Individualism* (ECPR Press, 2006).

2: EVERYTHING EVOLVES

1 Tom Robbins, *Still Life with Woodpecker* (Sidgwick & Jackson, 1980) p . 19.

2 Part of the reason why big animals get smaller on islands is that there is a limit to the amount of energy available to the population. The transition from small to big may have more to do with *intra-specific* competition than with competition between species.

3 For a good overview, see Thomas J. Matthews and Kostas Triantis, 'Island Biogeography', *Current Biology* 31: 19, 2021, R1201–R1207.

4 See Matt Ridley, *How Innovation Works: Serendipity, Energy and the Saving of Time* (Fourth Estate, 2020); *The Evolution of Everything: How Ideas Emerge* (Fourth Estate, 2015).

5 This, Ridley says, is morally the opposite of Social Darwinism (the idea that we should effect progress in society by interfering with human reproduction) because it means that ideas, not humans, compete and die.

6 See Nikolai A. Shevchuk, 'Adapted Cold Shower as a Potential Treatment for Depression', *Medical Hypotheses* 70: 5, 2008, pp. 995–1001.

7 Geert A. Buijze, Inger N. Sierevelt, Bas C. J. M. van der Heijden, Marcel G. Dijkgraaf and Monique H. W. Frings-Dresen, 'The Effect of Cold Showering on Health and Work: A Randomized Controlled Trial', *PLoS ONE* 11: 9, 2016, e0161749. Erratum in *PLoS ONE* 13: 8, 2018, e0201978. Knechtle Beat, Zbigniew Waśkiewicz, Caio Victor Sousa, Lee Hill and Pantelis T. Nikolaidis, 'Cold Water Swimming – Benefits and Risks: A Narrative Review', *International Journal of Environmental Research and Public Health* 17: 23, 2020, p. 8984; John S. Kelly and Ellis Bird, 'Improved Mood Following a Single Immersion in Cold Water', *Lifestyle Medicine* 3: 1, 2022, e53; Christoffer Van Tulleken, Michael Tipton, Heather Massey and C. Mark Harper, 'Open Water Swimming as a

Treatment for Major Depressive Disorder', *BMJ Case Reports* 2018, 2018: bcr-2018-225007.

8   A. Mooventhan and L. Nivethitha, 'Scientific Evidence-Based Effects of Hydrotherapy on Various Systems of the Body', '*North American Journal of Medical Sciences* 6: 5, May 2014, pp. 199–209.

9   Rhonda P. Patrick and Teresa L. Johnson, 'Sauna Use as a Lifestyle Practice to Extend Healthspan', *Experimental Gerontology* 154, 2021, p. 111509.

10 See E. J. Calabrese, 'Hormesis: A Fundamental Concept in Biology', *Microbial Cell* 1: 5, 2014, pp. 145–9; Suresh I. S. Rattan, 'Hormesis in Aging', *Ageing Research Reviews* 7: 1, 2008, pp. 63–78; David Gems and Linda Partridge, 'Stress-Response Hormesis and Aging: "That Which Does Not Kill Us Makes Us Stronger"', *Cell Metabolism* 7: 3, 2008, pp. 200–3.

11 S. Sutou, 'Low-Dose Radiation from A-Bombs Elongated Lifespan and Reduced Cancer Mortality Relative to Un-irradiated Individuals', *Genes and Environment* 40: 1, 2018, p. 26.

12 Edward J. Calabrese and Linda A. Baldwin, 'The Frequency of U-Shaped Dose Responses in the Toxicological Literature', *Toxicological Sciences* 62: 2, 2001, pp. 330–8.

13 This is just one scenario. The phenomenon of antibiotic resistance is complex. For an overview, see Marianne Frieri, Krishan Kumar and Anthony Boutin, 'Antibiotic Resistance', *Journal of Infection and Public Health* 10: 4, 2017, pp. 369–78.

14 For discussion, see Stephen C. Stearns, *The Evolution of Sex and Its Consequences*, vol. 55 (Birkhäuser, 2013); Matthew Hartfield and Peter D. Keightley, 'Current Hypotheses for the Evolution of Sex and Recombination', *Integrative Zoology* 7: 2, 2012, pp. 192–209; John Maynard Smith, *The Evolution of Sex*, vol. 4 (Cambridge University Press, 1978).

15 The tale of Pete and Maria is based on a true story.

16 Y. Apostolopoulos, S. Sönmez and C. H. Yu, 'HIV-Risk Behaviors of American Spring Break Vacationers: A Case of Situational Disinhibition?' *International Journal of STD & AIDS* 13, 2002, pp. 733–43.

17 A. Matteeli and G. Carosi, 'Sexually Transmitted Diseases in Travelers', *Clinical Infectious Disease* 32, 2001, pp. 1063–7.

18 For a good overview, see Michelle R. Kaufman, Andrea R. Fuhrel-Forbis, Seth C. Kalichman, Lisa A. Eaton, Demetria Cain, Charsey Cherry and Howard L. Pope, 'On Holiday: A Risk Behavior Profile for Men Who Have Vacationed at Gay Resorts', *Journal of Homosexuality* 56: 8, 2009, pp. 1134–44.

19 K. Wellings, W. Macdowall, M. Catchpole and J. Goodrich, 'Seasonal Variations in Sexual Activity and Their Implications for Sexual Health Promotion', *Journal of the Royal Society of Medicine* 92, 1999, pp. 60–4.

20 D. K. Whittier, J. S. St Lawrence and S. Seeley, 'Sexual Risk Behavior of Men Who Have Sex with Men: Comparison of Behavior at Home and at a Gay Resort', *Archives of Sexual Behaviour* 34, 2005, pp. 95–102.

21 Eric G. Benotsch et al., 'Sexual Risk Behavior in Men Attending Mardi Gras Celebrations in New Orleans, Louisiana', *Journal of Community Health* 32: 5, October 2007, pp. 343–56.

22 A. Anderson, S. Chilczuk, K. Nelson, R. Ruther and C. Wall-Scheffler, 'The Myth of Man the Hunter: Women's Contribution to the Hunt Across Ethnographic Contexts', *PLoS ONE* 18: 6, 2023, e0287101. Accessed at: https://doi.org/10.1371/journal.pone.0287101.

23 K. Hill and H. Kaplan, 'Trade-Offs in Male and Female Reproductive Strategies Among the Ache', in *Human Reproductive Behaviour*, ed. L. Batzig, M. Borgerhof Mulder and P. Turke (Cambridge University Press, 1988), pp. 272–90.

24 For discussion, see Geoffrey I. Crouch, 'Homo Sapiens on Vacation: What Can We Learn from Darwin?' *Journal of Travel Research* 52: 5, 2013, pp. 575–90.

25 Ibid, citing T. Flannery, *The Weather Makers* (Grove Press, 2005).

3: A HUMAN AT THE EDGE OF EUROPE

1 Nelleke IJssennagger, *Central because Liminal: Frisia in a Viking Age North Sea World* , PhD diss., University of Groningen, 2017, p. 256.

2 Eppie R. Jones, Gloria Gonzalez-Fortes, Sarah Connell, Veronika Siska, Anders Eriksson, Rui Martiniano, Russell L. McLaughlin et al., 'Upper Palaeolithic Genomes Reveal Deep Roots of Modern Eurasians', *Nature Communications* 6: 1, 2015, p. 8912.

4: CAN ANYTHING GOOD COME FROM A CITY?

1 Lewis Mumford, *The City in History: Its Origins, its Transformations and Its Prospects* (Harcourt, Brace, 1961), p. 559.

2 E. M. Forster, *Howards End* (Edward Arnold, 1910), p. 147.

3 There is a vast and forbidding literature on the birth of cities. The most accessible summary I know is in the podcast by Tom Holland and Dominic Sandbrook, mainly about Uruk, entitled 'The World's First City', www.youtube.com/watch?v=s7u6p9VDl-Y.

4 Leopold Kohr, *The Breakdown of Nations* (Bloomsbury, 2017).

5 There are many obvious omissions from this list. The absence of any-thing non-Western (with the exception of Byzantium) would be culpable if I were trying to do anything other than exemplify. China, India, Meso-potamia, Egypt, Carthage, the great cities of the Levant and the ancient and modern cultures of the New World (along with many others) would need to feature in any work that claimed to be a systematic exposition of the thesis. I address the case of Jerusalem in Chapters 6 and 11.

6 Rex Warner and Martin Hurlimann, *Eternal Greece* (Readers Union, 1955), p. 86.

7 Andrew Graham-Dixon, *Renaissance* (BBC Books, 1999).

8 A turn violently resented by Savonarola, who did his energetic best to re-sacralize Florence.

9 William Blake, *Jerusalem: The Emanation of the Giant Albion* (William Blake, self-published, 1804).

10 Charles Dickens, *Bleak House* (Bradbury and Evans, 1853), pp. 1–2.

11 Tony Judt, 'Edge People', *The New York Review of Books*, 25 March 2010. Accessed at: www.nybooks.com/articles/2010/03/25/edge-people.

12 Nassim Nicholas Taleb, *The Black Swan: The Impact of the Highly Improb-able* (Random House, 2010).

13 For discussion of the various ways in which Bacon expressed this idea, see Eugene P. McCreary, 'Bacon's Theory of Imagination Reconsidered', *Huntington Library Quarterly* 36: 4, 1973, pp. 317–26.

14 M. Lichtheim, *Ancient Egyptian Autobiographies Chiefly of the Middle Kingdom: A Study and an Anthology* (Universitätsverlag, 1988), pp. 25–6. Ankhtifi conquered Edfu and claimed to control Aswan. He was loyal to the kings of the new dynasty established during the First Intermedi-ate Period in Heracleopolis Magna, about 90 km south of the old seat of power, Memphis.

15 'The First Intermediate Period (*c.* 2160–2055 BC)', in *The Oxford History of Ancient Egypt*, ed. Ian Shaw Oxford University Press, (2000), p. 127.

16 A reflection of the increased interest in peripheral contributors to artis-tic projects is the recent move from auteur theory to an increased focus on (for instance) producers and technicians in the film industry. For discussion, see Andrew Butterfield's review of James Hall's *The Artist's Studio: A Cultural History*: A. Butterfield, 'Pop Goes the Easel', *Times Literary Supplement*, 17 February 2023.

17 See too W. Ejsmond, 'The First Intermediate Period in Egypt', *Oxford Research Encyclopedia of African History*, 17 July 2024. Accessed at:

https://oxfordre.com/africanhistory/view/10.1093/acrefore/9780190277734
.001.0001/acrefore-9780190277734-e-1203.

5: THE MACHINE STATE

1 Constantine Cavafy, 'The Satrapy', trans. John Stathatos (unpublished, 2024). Accessed at: www.stathatos.net/texts/translations/satrapy. For a comment on this poem, see p. 217.
2 P. J. O'Rourke, *Holidays in Hell* (Atlantic Monthly Press, 1988), p. xvi. The section in square brackets is my interpolation.
3 E. M. Forster, *Howards End* (Edward Arnold, 1910), p. 158.
4 Paul Kingsnorth, 'The Jellyfish Tribe', *The Abbey of Misrule*, Substack, 16 March 2023.
5 James C. Scott, *The Art of Not Being Governed: An Anarchist History of Upland Southeast Asia* (Yale University Press, 2009).
6 Lois Beck, 'Tribes and the State in 19th- and 20th-Century Iran', in *Tribes and State Formation*, eds Philip Koury and Joseph Kostiner (University of California Press, 1990), pp. 191 and 192, cited in ibid, p. 210.
7 Most law enforcement doesn't come from the Supreme Court, but from the pressure of your conscience, the expectations of your community, the police officer whose kids go to the same school as yours, and the lay magistrates whose local knowledge, parochial concern and instinct for fairness usually triumph over the fine distinctions drawn by bewigged judges and ivory-towered academics. In England and Wales magistrates, when they're sworn in, take this oath: 'I . . . do swear by Almighty God that I will well and truly serve our Sovereign Lord King Charles the Third in the office of justice of the peace, and I will do right to all manner of people after the laws and usages of this realm, without fear or favour, affection or ill will.' When they're delivering their local justice (perhaps rough, approximate justice by the standards of the Supreme Court), magistrates are generally doing their best to honour that oath. They humanize, localize and contextualize the law.
8 Kingsnorth, 'The Jellyfish Tribe'.
9 Noam Chomsky, *The Umbrella of U.S. Power* (Seven Stories Press, 2002), p. 42. The whole sentence from which this is taken reads: 'States are not moral agents; people are, and can impose moral standards on powerful institutions.'

6: TOUCHING THE EDGE OF GOD'S ROBE

1 See David Lewis-Williams, 'Religion and Archaeology: An Analytical, Materialist Account', in *Belief in the Past: Theoretical Approaches to the Archaeology of Religion*, ed. David S. Whitley (Routledge, 2015), pp. 23–42; David Lewis-Williams and Sam Challis, *Deciphering Ancient Minds: The Mystery of San Bushman Rock Art* (Thames & Hudson, 2012); David Lewis-Williams, *The Mind in the Cave: Consciousness and the Origins of Art* (Thames & Hudson, 2011).

2 Laurens van der Post, *The Lost World of the Kalahari* (Hogarth Press, 1958). For a general discussion of telepathy, including mention of the Laurens van de Post example, see Jeremy Smith, 'Telepathy: A New Way of Seeing', *The Ecologist*, 1 September 2005. Accessed at: https://theecologist.org/2005/sep/01/telepathy-new-way-seeing. For a detailed discussion of telepathy in non-human animals, see Rupert Sheldrake, *Dogs That Know Their Owners Are Coming Home: And Other Unexplained Powers of Animals* (Crown, 2011).

3 Quran 27:15–44. The Jewish sources include Targum Sheni, an Aramaic paraphrase of the book of Esther.

4 Genesis 12:10–20.

5 Genesis 23.

6 The basis of the rules of *kashrut* is found in Leviticus 11:3–12 and Deuteronomy 14:4–10.

7 I am grateful to Nicholas Stone for this observation.

8 John 1:46.

9 Matthew 8:20.

10 Blaise Pascal, *Pascal's Pensées* (E. P. Dutton & Co, 1958), p. 233, entry 785.

11 Matthew 6:20.

12 C. S. Lewis, 'The Weight of Glory', in *Screwtape Proposes a Toast* (Fontana, 1965).

13 The current population of Mecca is around 2.4 million. There are estimated to be 1.9 billion Muslims in the world.

14 John Bunyan, *The Pilgrim's Progress from This World to That Which is to Come* (1678).

15 For a wide-ranging general examination of this proposition, see *Religion on the Edge: De-centering and Re-centering the Sociology of Religion*, ed. Courtney Bender (Oxford University Press, 2013).

16 Flinders Petrie, *Tell el-Amarna* (Methuen, 1894), p. 42.

7: STATES OF MIND AND UN-MIND

1 Khalil Gibran, *The Madman: His Parables and Poems* (Knopf, 1918), p. 11.

2 Tim Hughes, 'Consuming Fire' (Survivor Records, 2004).

3 I give a fuller account of this experience in *Being a Human* (Profile, 2021).

4 Charles Dickens, *Oliver Twist* (Wordsworth Classics, 1992), p. 222.

5 Célia Lacaux, Thomas Andrillon, Céleste Bastoul, Yannis Idir, Alexandrine Fonteix-Galet, Isabelle Arnulf and Delphine Oudiette, 'Sleep Onset is a Creative Sweet Spot', *Science Advances* 7: 50, 2021, eabj5866.

6 For general discussion of sleep deprivation in religious practice, see Núria M. Farré-i-Barril, 'Sleep Deprivation: Asceticism, Religious Experience and Neurological Quandaries', in *Religion and the Body*, eds David Cave and Rebecca Sachs Norris (Brill, 2012), pp. 217–34; and Sarah Macmillan, ' "The Nyghtes Watchys": Sleep Deprivation in Medieval Devotional Culture', *Journal of Medieval Religious Cultures* 39: 1, 2013, pp. 23–42.

7 Mother Teresa, *Come Be My Light: The Private Writings of the 'Saint of Calcutta'*, edited and with commentary by Brian Kolodiejchuk (Doubleday, 2007), p. 262.

8 I am referring to the experience known as the 'dark night of the soul', familiar to many of the great mystics. The classic example, of course, is St John of the Cross.

9 Vilayat Inayat Khan, *Thinking Like the Universe: The Sufi Path of Awakening* (Thorsons, 2000), p. 29.

10 It has never been coherently (nor even very enthusiastically) suggested that consciousness, in the sense of subjectivity, would be visible to natural selection. Of course natural selection notices and selects vigorously for anything that promotes relationality and community (since those indeed can confer a colossal selective advantage), but you don't need subjectivity for that. If anything, subjectivity might get in the way, making us examine our navels rather than curate our relationships.

11 For detailed discussion, see Andrew McGee and Charles Foster, *Intuitively Rational: How We Think and How We Should* (Springer, 2024).

12 Other epileptics in Dostoyevsky's work include Smerdyakov in *The Brothers Karamazov*. Dostoyevsky's fictional epileptics are generally understood to reflect his own experiences. For further discussion, see I. Iniesta, 'Epilepsy in the Process of Artistic Creation of Dostoevsky', *Neurología* (English Edition) 29: 6, 2014, pp. 371–8.

13 For detailed discussion, see Charles Foster, *Wired for God: The Biology of Spiritual Experience* (Hodder & Stoughton, 2011).

14 Andy Mitchell, *Ten Trips: The New Reality of Psychedelics* (The Bodley Head, 2023), p. 222.

15 Michael W. Reimann, Max Nolte, Martina Scolamiero, Katharine Turner, Rodrigo Perin, Giuseppe Chindemi, Paweł Dłotko, Ran Levi, Kathryn Hess and Henry Markram, 'Cliques of Neurons Bound into Cavities Provide a Missing Link Between Structure and Function', *Frontiers in Computational Neuroscience* 11, 2017, p. 48. For more general discussion, see Christian Bick, Elizabeth Gross, Heather A. Harrington and Michael T. Schaub, 'What are Higher-Order Networks?' *SIAM Review* 65: 3, 2023, pp. 686–731; and C. W. Lynn and D. S. Bassett, 'The Physics of Brain Network Structure, Function and Control', *Nature Reviews Physics* 1, 2019, pp. 318–32.

8: DESTITUTION AND DEATH

1 From the album *Lonesome Troubadour* (1991).

2 Roberta Davidson, 'Prison and Knightly Identity in Sir Thomas Malory's Morte D'Arthur', *Arthuriana* 14: 2, 2004, pp. 54–63, 61.

3 There are many modern glosses. See, for instance, Jussi Suikkanen, 'An Improved Whole Life Satisfaction Theory of Happiness', *International Journal of Wellbeing* 1: 1, 2011, pp. 149–66. For general discussion, see Guy Fletcher, *The Philosophy of Well-Being: An Introduction* (Routledge, 2016); and Lorraine L. Besser, *The Philosophy of Happiness: An Interdisciplinary Introduction* (Routledge, 2020).

4 Aristotle, *Nicomachean Ethics* 1097b22–1098a18. See the discussion in Anthony Kenny, *Aristotle and the Perfect Life* (Oxford University Press, 1992). The inflection in the *Eudemian Ethics* is slightly different: there, happiness is described as the activity of a perfect life in accordance with perfect virtue. For a modern commentary, see Jonathan Phillips, Sven Nyholm and Shen-yi Liao, 'The Good in Happiness', in *Oxford Studies in Experimental Philosophy: Volume 1* (Oxford University Press, 2014), pp. 253–93. The influence on me of Alasdair MacIntyre's work (and particularly *After Virtue: A Study in Moral Theory* (University of Notre Dame Press, 1981) and *Dependent Rational Animals: Why Human Beings Need the Virtues* (Open Court, 1999) is obvious and profound.

5 Another good example is Dostoyevsky. In 1849, at the age of twenty-eight, he was led to a stake in a public square in St Petersburg. His sentence was read aloud: he was to be executed by firing squad. Moments from death,

he kissed a cross. But it was all a Tsarist stunt. He was reprieved at the last moment and sent instead to a Siberian labour camp. The experience changed him for ever. Shortly after his reprieve, he wrote to his brother, Mikhail: '[T]o be a *human being* among people and to remain one forever, no matter in what circumstances, not to grow despondent and not to lose heart – that's what life is all about, that's its task . . . If anyone remembers me with malice, and if I quarrelled with anyone, if I made a bad impression on anyone – tell them to forget about that if you manage to see them. There is no bile or spite in my soul, I would like to so love and embrace at least someone out of the past at this moment . . . When I look back at the past and think how much time was spent in vain, how much of it was lost in delusions, in errors, in idleness, in the inability to live; how I failed to value it, how many times I sinned against my heart and spirit – then my heart contracts in pain. Life is a gift, life is happiness, each moment could have been an eternity of happiness. Si jeunesse savait! . . . Now, changing my life, I'm being regenerated into a new form. Brother! I swear to you that I won't lose hope and will preserve my heart and spirit in purity. I'll be reborn for the better. That's my entire hope, my entire consolation. Life in the casemate has already sufficiently killed off in me the needs of the flesh that were not completely pure . . . ' Fyodor Dostoevsky, *Dostoevsky: Letters*, ed. David Lowe and Ronald Meyer (Ardis, 1988) (emphasis in original).

6 I am thinking, of course, of Joseph Conrad's *Heart of Darkness*, published as a serial by *Blackwood's Magazine*, 1899, and as a book by William Blackwood and Sons, 1902. Cf. Edwin M. Lemert, *The Trouble with Evil: Social Control at the Edge of Morality* (State University of New York Press, 1997).

7 David Bentley Hart, *The New Testament: A Translation* (Yale University Press, 2019).

8 'Entering the Moral Maze: A Discussion on Religion and Ethics with Richard Holloway', *Three Monkeys Online*. Accessed at: https://www. threemonkeysonline.com/entering-the-moral-maze-a-discussion-on-religion-and-ethics-with-richard-holloway.

9 Jay Griffiths, *Why Rebel?* (Penguin, 2021), p. 39.

10 E. M. Forster, *Howards End* (Edward Arnold, 1910), p. 349.

9: DOGS ON CORNFLAKE PACKETS

1 John Steinbeck, *Journal of a Novel: The East of Eden Letters* (Pan, 1977), p. 4.

2 Claudia Oik, *Shakespeare and Beckett: Restless Echoes* (Cambridge University Press, 2023), p. 12.

3 A. N. Whitehead, *American Essays in Social Philosophy* (Greenwood Publishing Group, 1975).

4 Anaïs Nin, *The Diary of Anaïs Nin, Volume 3: 1939–1944*, ed Gunther Stuhlman (Mariner Books, 1971), entry for October 1943.

5 Source unknown. Friedrich (1774–1840) was a German landscape painter.

6 Jay Griffiths, *Why Rebel?* p. 37.

7 C. S. Lewis, *The Discarded Image* (Cambridge University Press, 1964).

8 Virginia Woolf, *Moments of Being*, ed. Jeanne Schulkind (Mariner Books, 1985), p. 81.

9 Eric Wargo, *From Nowhere: Artists, Writers, and the Precognitive Imagination* (Anomalist Books, 2024).

10 George and Weedon Grossmith, *Diary of a Nobody* (J. W. Arrowsmith, 1892).

11 John Buchan, *John Macnab* (Hodder & Stoughton, 1925).

12 *Sir Gawain and the Green Knight*, trans. J. R. R. Tolkien (Allen & Unwin, 1979), p. 15.

13 Ibid, p. 16.

14 Ibid, pp. 32–3.

15 Ibid, p. 32.

16 Gillian Rudd, 'The Wilderness of Wirral in *Sir Gawain and the Green Knight*', *Arthuriana* 23: 1, 2013, pp. 52–65.

17 Emphasis added.

18 James Cahill, *Tiepolo Blue* (Hachette, 2022), p. 30.

19 Cited in Rex Warner and Martin Hurlimann, *Eternal Greece* (Readers Union, 1955), p. 9.

20 Cited in ibid, p. 36.

21 *Blake's Exhibition and Catalogue of 1809*, pp. 63–4.

10: WHAT ARE WE? WHERE ARE WE?

1 Francesco Petrarch, *Familiar Letters*, vol. 1, trans. Morris Bishop (Indiana University Press, 1963), 4.1.

2 Psalm 8:4–5.

3 At various points in this book I suggest that we can draw some conclusions about the sorts of creatures we constitutionally are – and, by extension, how we should behave – based on certain facts about ourselves and our evolutionary history.

The second part of that suggestion (concerning how we should behave) will raise the suspicions of anyone who has ever studied philosophy, because it appears to embody the naturalistic fallacy – the heresy of deriving an 'ought' from an 'is'.

Exactly what Hume's famous claim about 'is' and 'ought' entails is bitterly contested and exhaustingly discussed. This whole area of philosophy is littered with the corpses of straw men.

What does it mean? Possibly many things, but it means at least this: that one can infer an *ought* from an *is* only if, in addition to matters of mere fact, one includes an evaluative premise.

I don't doubt the utility of this principle. In some situations it is very useful. But its utility depends on the context. Where there is overwhelming consensus about the relevant evaluative premise, one can leave the 'is–ought' distinction in the tool bag. Where no such consensus exists, one needs to take it out and use it.

I hope that when I do try to derive conclusions about how we should behave, there is indeed overwhelming consensus about the relevant evaluative principle.

For an accessible account of the 'is–ought' gap, see the entry on Moral Naturalism in the *Stanford Encyclopaedia of Philosophy*: https://plato.stanford.edu. I am grateful to commentators on a post I wrote on the subject for the University of Oxford's *Practical Ethics* blog.

4 Freya Stark, *Perseus in the Wind: A Life of Travel* (John Murray, 1948).

5 In *The Better Angels of Our Nature* (Viking, 2011), Steven Pinker argues that violence has declined over long stretches of human history. I discuss his thesis in detail in *Being a Human* (Profile, 2021).

6 See David Lewis-Williams, 'Religion and Archaeology: An Analytical, Materialist Account', in *Belief in the Past* (Routledge, 2016), pp. 23–42; David Lewis-Williams and Sam Challis, *Deciphering Ancient Minds: The Mystery of San Bushman Rock Art* (Thames & Hudson, 2012); David Lewis-Williams, *The Mind in the Cave: Consciousness and the Origins of Art* (Thames & Hudson, 2011); and David Lewis-Williams, E. Thomas Lawson, Knut Helskog, David S. Whitley and Paul Mellars, 'Review Feature: A Review of *The Mind in the Cave: Consciousness and the Origins of Art* by David Lewis-Williams', *Cambridge Archaeological Journal* 13: 2, 2003, pp. 263–79. See also David Lewis-Williams, 'Harnessing the Brain: Vision and Shamanism in Upper Palaeolithic Western Europe', *Beyond Art: Pleistocene Image and Symbol* 23, 1997, pp. 321–42. Lewis-Williams's view has been robustly criticized. See, for instance, Grant S. McCall, 'Add Shamans and Stir? A Critical Review of the Shamanism Model of Forager Rock Art Production', *Journal of Anthropological Archaeology* 26:

2, 2007, pp. 224–33; and Richard Bradley, *Image and Audience: Rethinking Prehistoric Art* (Oxford University Press, 2009). Few today would contend that Lewis-Williams's account is the whole story.

7 Alan Garner, 'Achilles in Altjira', in *The Voice that Thunders* (Harvill, 2010), p. 58.

8 Paul Pettitt, 'Landscapes of the Dead: The Evolution of Human Mortuary Activity from Body to Place in Palaeolithic Europe' in *Settlement, Society and Cognition in Human Evolution: Landscape in Mind*, eds F. Coward, R. Hosfield, M. Pope and F. Wenban-Smith (Cambridge University Press, 2015), pp. 258–74, 263. See also P. Bloom, 'Religion Is Natural', *Developmental Science* 10: 1, 2007, pp. 147–51, 148.

9 I am thinking, of course, of the *Löwenmensch*, or Lion-Man, found at Hohlenstein-Stadel in the Swabian Jura. It is dated to between 35,000 and 41,000 years ago.

10 Martha C. Nussbaum, *Creating Capabilities: The Human Development Approach* (Harvard University Press, 2011).

11 Matthew 5:3.

12 See Luke 12:16–21.

13 The notions of quantum non-locality and entanglement.

14 Iain McGilchrist, *The Matter with Things: Our Brains, Our Delusions and the Unmaking of the World* (Perspectiva, 2022).

15 C. S. Lewis, *The Voyage of the Dawn Treader* (Geoffrey Bles, 1952).

16 *Savoir* and *wissen* connote 'head' knowledge; *connaître* and *kennen* familiarity.

17 McGilchrist, *The Matter with Things*.

18 McGilchrist himself, following the eighteenth- and nineteenth-century German philosopher Friedrich Schelling, prefers the metaphor of the whirlpool, which illustrates well the idea – crucial to McGilchrist's thesis – that resistance, arising from the flow itself, is part of the creative force of the cosmos: it is the resistance that enables the unfolding of its potential. Schelling observes: 'Where there is resistance – a whirlpool forms. Every original product of nature is such a whirlpool, every organism. The whirlpool is not something immobilized, it is rather something constantly transforming – but reproduced anew at each moment. Thus no product in nature is *fixed*, but it is reproduced at each instant through the force of nature entire . . . Nature as a whole cooperates in every product . . . ' Friedrich Wilhelm Joseph von Schelling, *First Outline of a System of the Philosophy of Nature* (SUNY Press, 2004), 'The Unconditioned in Nature', vol. 1, i, p. 18.

19 See, for example, the discussion in McGilchrist's *The Matter with Things*, pp. 945–96.

20 Jenny Odell, *Saving Time: Discovering a Life Beyond the Clock* (Vintage, 2024).

21 William James observed that all the entities in the universe 'interdigitate with their next neighbours in manifold directions, and there are no clean cuts between them anywhere' (cited in McGilchrist's *The Matter with Things*, p. 945). This would pose a challenge to my thesis were it not for the corollary that, time behaving as it does, humans – and everything else – are constantly on the wild frontier of the unrolling cosmos.

22 'Bishop Blougram's Apology' is a long poem in Robert Browning's 1855 collection, *Men and Women*.

23 Maximus also speaks of humans as *mediators*, tasked with the mission of bridging the edges – mediating between individuals and fostering the relationships that are of the essence of all things.

II: VIEWPOINTS

1 Andrew Louth, *Modern Orthodox Thinkers: From the Philokalia to the Present* (SPCK, 2015), pp. 282–3.

2 N. H. Reeve and Richard Kerridge, *Nearly Too Much: The Poetry of J. H. Prynne* (Liverpool University Press, 1995).

3 Freya Stark agrees: 'The tourist travels in his own atmosphere like a snail in his shell and stands, as it were, on his own perambulating doorstep to look at the continents of the world. But if you discard all this, and sally forth with a leisurely and blank mind, there is no knowing what may not happen to you.' *Baghdad Sketches: Journeys Through Iraq* (John Murray, 1937).

4 James Cahill, *Tiepolo Blue* (Hachette, 2022).

5 Rudyard Kipling, 'The English Flag', in *The Seven Seas* (Methuen, 1896).

6 See, for example, Avraham Faust, *Israel's Ethnogenesis: Settlement, Interaction, Expansion and Resistance* (Routledge, 2016); Jacob L. Wright, *War, Memory, and National Identity in the Hebrew Bible* (Cambridge University Press, 2020); William G. Dever, 'Archaeology, Urbanism, and the Rise of the Israelite State', *Journal for the Study of the Old Testament Supplement Series* 1997, pp. 172–93.

7 We saw in Chapter 6 some senses in which it is not – in which it is contingent, conditional: a tent; a work in progress.

8 Compare Blake's vision of 'eternity in a grain of sand', where he suggests that the infinite can be seen in the infinitesimal. See 'Auguries

of Innocence', written *c.* 1803, published posthumously in *Poetical Sketches* and later included in *The Pickering Manuscript* (Pickering, 1863).

9 Chiang Lee, *The Silent Traveller in Oxford* (Methuen, 1944). He wrote similarly acute books about the Lake District, London, the Yorkshire Dales, Edinburgh, New York, Dublin, Paris, Boston, San Francisco and Japan.

10 Madeleine Bunting, *The Seaside: England's Love Affair* (Granta, 2023).

11 Nigel McGilchrist, *When the Dog Speaks, the Philosopher Listens* (Genius Loci, 2023).

12 A land area of 131,957 km² divided by the length of coastline (13,676 km).

13 547,030 km² divided by 3,427 km.

14 See, for instance, Kay Redfield Jamison, 'Manic-Depressive Illness and Creativity', *Scientific American* 272: 2, 1995, pp. 62–7; and Shelley H. Carson, 'Creativity and Mental Illness', in *The Cambridge Handbook of Creativity* (Cambridge University Press, 2019), pp. 296–318. There is, however, great individual variability, and in terms of *population* effects the connection between psychopathology and creativity is weak or non-existent. See the meta-analysis by Sue Hyeon Paek, Ahmed M. Abdulla and Bonnie Cramond, 'A Meta-Analysis of the Relationship Between Three Common Psychopathologies – ADHD, anxiety, and depression – and Indicators of Little-c Creativity', *Gifted Child Quarterly* 60: 2, 2016, pp. 117–33.

15 *The Notebooks of Leonardo da Vinci*, trans. Jean Paul Richter (Dover Publications, 1970), entry 295.

16 The early beach scenes painted by Salvador Dalí show Dalí alone on the beach, too, staring out at the sea.

17 Alex Pilcher, commenting on *Repose on the Flight from Egypt*, says: 'Hybrid creatures often stalk the realms of queer imagination. Philpot's mentor Charles Ricketts (1866–1931) illustrated a lavish edition of Oscar Wilde's coded, erotic poem *The Sphinx*. Vaslav Nijinsky became a queer icon through his performance as a faun. Centaurs and minotaurs still crop up both in gay erotic comics and in queer contemporary art. Such creatures of the wilds embody the promise of sensuality untrammelled by human social conventions. During the Italian Renaissance of the fifteenth century, these sexy, classically-inspired hybrids returned to the stage of Western art, coinciding with a rediscovery of the idealised male nude.' *A Queer Little History of Art* (Tate Publishing, 2017).

18 The full passage reads: 'I should not like to close without attempting to set before you, though only in dim outline, the ideal towards which poetic drama should strive. It is an unattainable ideal: and that is why it interests me, for it provides an incentive towards further experiment and exploration,

beyond any goal which there is prospect of attaining. It is a function of all art to give us some perception of an order in life, by imposing an order upon it. The painter works by selection, combination and emphasis among the elements of the visible world; the musician in the world of sound. It seems to me that beyond the namable, classifiable emotions and motives of our conscious life when directed towards action – the part of life which prose drama is wholly adequate to express – there is a fringe of indefinite extent, of feeling which we can only detect, so to speak, out of the corner of the eye and can never completely focus; of feeling of which we are only aware in a kind of temporary detachment from action. There are great prose dramatists – such as Ibsen and Chekhov – who have at times done things of which I would not otherwise have supposed prose to be capable, but who seem to me, in spite of their success, to have been hampered in expression by writing in prose. This peculiar range of sensibility can be expressed by dramatic poetry, at its moments of greatest intensity. At such moments, we touch the border of those feelings which only music can express.' T. S. Eliot, *Poetry and Drama* (Harvard University Press, 1951), p. 42.

See too this letter from the poet Elizabeth Bishop to Anne Stevenson, dated 8 January 1964: 'Yes, I agree with you. I think that's what I was trying to say in the speech above. There is no "split." Dreams, works of art (some), glimpses of the always-more-successful surrealism of everyday life, unexpected moments of empathy (is it?), catch a peripheral vision of whatever it is one can never really see full-face but that seems enormously important. I can't believe we are wholly irrational – and I do admire Darwin! But reading Darwin, one admires the beautiful and solid case being built up out of his endless heroic observations, almost unconscious or automatic – and then comes a sudden relaxation, a forgetful phrase, and one feels the strangeness of his undertaking, sees the lonely young man, his eyes fixed on facts and minute details, sinking or sliding giddily off into the unknown. What one seems to want in art, in experiencing it, is the same thing that is necessary for its creation, a self-forgetful, perfectly useless concentration.' Elizabeth Bishop, *Prose: Centenary Edition* (Chatto & Windus, 2011), p. 414.

19 In a 1931 lecture on mathematical semantics to the American Mathematical Society. See http://esgs.free.fr/uk/art/sands-sup3.pdf.

20 McGilchrist puts it arrestingly: the re-presentation to the right hemisphere results in *com-prehension* – the coming together of disparate elements to make a whole. The left hemisphere's contribution is *ap-prehension* – grasping.

21 William Blake, *The Marriage of Heaven and Hell* (1790–3).

12: THE GREAT PROCESSION

1 The quote continues: '. . . the light gleams an instant, then it's night once more.' Samuel Beckett, *Waiting for Godot* (1955), Act 2.

2 Mishnah, Kelim 1:1 and Babylonian Talmud (Avodah Zarah) 37b; cf. Numbers 31:19 and 19:22.

3 See 'David Fuller: killer who abused mortuary bodies will die in jail', BBC News, 15 December 2021. Accessed at: https://www.bbc.co.uk/news/uk-england-kent-59601656.

4 'The Keys To The Underworld' in *Greek Folk Songs*, trans. Joshua Barley (Aiora Press, 2022), p. 157.

5 Luke 18:17.

6 Thomas Traherne, *Centuries of Meditations*, ed. Michael L. Byrd (D. S. Brewer, 2002).

7 Wordsworth's *Prelude* (1850) is the obvious example.

8 Peter Matthiessen, *At Play in the Fields of the Lord* (Random House, 1965).

13: MARKING THE EDGES

1 Sir Thomas Browne, *Hydriotaphia, or Urn Burial* (1658).

2 For an account of Gawain's self-realization in the liminal space of the Wilderness of Wirral, see Gillian Rudd, 'The Wilderness of Wirral, in *Sir Gawain and the Green Knight*', *Arthuriana* 23: 1, 2013, pp. 52–65.

3 Dante Alighieri, *Inferno* 1:10. There are, of course, hundreds of translations. This, so far as I can ascertain, is my own gloss.

4 *Sir Gawain and the Green Knight*, trans. J. R. R. Tolkien (Allen & Unwin, 1975).

5 C. S. Lewis, *The Magician's Nephew* (The Bodley Head, 1955).

6 *Inferno* opens on Good Friday – the threshold of a liminal time in which Jesus, on a great adventure indeed, harrows Hell and builds a bridge between Earth and Heaven.

7 In Robert Graves, *The White Goddess: A Historical Grammar of Poetic Myth* (Faber & Faber, 1948).

8 In James George Frazer, *The Golden Bough* (Macmillan, 1890).

9 Revelation 21:1.

10 There are many other examples – such as the story of Lleu Llaw Gyffes, in *The Mabinogion*.

11 Victor Turner, *The Ritual Process: Structure and Anti-Structure* (Routledge & Kegan Paul, 1969), p. 146.

12 Tom Boylston, *The Shade of the Divine: Approaching the Sacred in an Ethiopian Orthodox Christian Community*, PhD diss. (London School of Economics, 2012), pp. 116–17. Accessed at: http://etheses.lse.ac.uk/339/1/Boylston_The%20Shade%20of%20the%20Divine.pdf

14: THE GREAT FRAUD

1 Nassim Nicholas Taleb, *The Black Swan: The Impact of the Highly Improbable* (Random House, 2010), p. 229.

2 Ibid, p. xxix.

3 For detailed discussion of this issue, and of the errors that creep in when we presume that laboratory conclusions about behaviour are meaningful, see Arik Kershenbaum, *Why Animals Talk* (Penguin, 2023).

4 There is another strong scientific reason to be suspicious of our current methods of assessing honours: mere luck is likely to be a more important determinant of 'success' than merit or effort. A 2018 paper observes: 'The largely dominant meritocratic paradigm of highly competitive Western cultures is rooted on the belief that success is mainly due, if not exclusively, to personal qualities such as talent, intelligence, skills, smartness, efforts, wilfulness, hard work or risk taking. Sometimes, we are willing to admit that a certain degree of luck could also play a role in achieving significant success. But, as a matter of fact, it is rather common to underestimate the importance of external forces in individual successful stories. It is very well known that intelligence (or, more in general, *talent* and personal qualities) exhibits a Gaussian distribution among the population, whereas the distribution of wealth – often considered as a proxy of success – follows typically a power law (Pareto law), with a large majority of poor people and a very small number of billionaires. Such a discrepancy between a Normal distribution of inputs, with a typical scale (the average talent or intelligence), and the scale-invariant distribution of outputs, suggests that some hidden ingredient is at work behind the scenes. In this paper, we suggest that such an ingredient is just randomness. In particular, our simple agent-based model shows that, if it is true that some degree of talent is necessary to be successful in life, almost never the most talented people reach the highest peaks of success, being overtaken by averagely talented but sensibly luckier individuals.' Alessandro Pluchino, Alessio Emanuele Biondo and Andrea Rapisarda, 'Talent Versus Luck: The Role of Randomness in Success and Failure', *Advances in Complex Systems* 21: 03n04, 2018, 1850014. I am aware that by citing this, I'm saying that randomness (the basic diet of the Gaussian bell curve) is useful in discussing human behaviour. That may seem to

chafe with my reliance on Taleb, but it does not. Taleb's black swans come out of nowhere – not out of a bell curve.

5   Graham Greene, *The Ministry of Fear* (Heinemann, 1943).

6   See https://www.margaretthatcher.org/document/106689.

7   In 2023. See https://www.statista.com/statistics/540836/adidas-marketing-spend/#:~:text=adidas%20Group%20POS%20marketing%20spend%20 2015%2D2023&text=The%20adidas%20Group%20incremented%20its, comparison%20to%20the%20previous%20year.

8   Life certainly can't be reduced to matter, not least because, as Joseph Needham noted, 'nothing can ever be reduced to anything.' See Joseph Needham, *Time: The Refreshing River (Essays and Addresses, 1932–1942)* (George Allen & Unwin, 1943), p. 183.

9   C. S. Lewis, *A Grief Observed* (Faber & Faber, 1961).

10 Andrew Harvey, *A Journey in Ladakh: Encounters with Buddhism* (Houghton Mifflin Harcourt, 2000).

11 Per Richard Feynman.

12 William Blake, 'Auguries of Innocence', written *c.* 1803, published posthumously in *Poetical Sketches*, and later included in *The Pickering Manuscript* (Pickering, 1863).

13 John Keats, letter to George and Thomas Keats, 21 December 1817, in *Selected Letters of John Keats*, ed. Robert Gittings (Oxford University Press, 2003), pp. 41–2.

14 Iain McGilchrist observes that Shakespeare shows '. . . complete disregard for theory and category, a celebration of multiplicity and the richness of human variety, rather than the rehearsal of common laws for personality and behaviour according to type. Shakespeare's characters are so stubbornly themselves, and not the thing that fate, or the dramatic plot, insists they should be . . . Richard II ill-suited to being king, more a self-absorbed poet; Macbeth overcome with scruples and visions of guilt, the reluctant usurper; Anthony, love-besotted, his will suborned, hardly the fearless military commander . . .' McGilchrist notes that Shakespeare confounded genres, placed comic scenes in tragedies, and confronted opposites – recognizing that the 'web of our life is of a mingled yarn, good and ill together.' Figures like Falstaff, he adds, are 'incompressible in terms of the elements into which they could be analysed, but form, Gestalt-like, new coherent, living wholes.' *The Master and His Emissary: The Divided Brain and the Making of the Western World* (Yale University Press, 2009), p. 304.

15: THE WAR ON THE EDGES

1 V. S. Naipaul, *Half a Life* (Knopf, 2001), p. 225.

2 Howard Jacobson, *J: A Novel* (Jonathan Cape, 2014).

3 Martha C. Nussbaum, *Upheavals of Thought: The Intelligence of Emotions* (Cambridge University Press, 2003). She says further: 'A lot is at stake in the decision to view emotions in this way, as intelligent responses to the perception of value. If emotions are suffused with intelligence and discernment, and if they contain in themselves an awareness of value or importance, they cannot, for example, easily be sidelined in accounts of ethical judgment, as so often they have been in the history of philosophy. Instead of viewing morality as a system of principles to be grasped by the detached intellect, and emotions as motivations that either support or subvert our choice to act according to principle, we will have to consider emotions as part and parcel of the system of ethical reasoning. We cannot plausibly omit them, once we acknowledge that emotions include in their content judgments that can be true or false, and good or bad guides to ethical choice. We will have to grapple with the messy material of grief and love, anger and fear, and the role these tumultuous experiences play in thought about the good and the just . . . '

4 See, for instance, David Butterfield, 'Hollowed-out Humanities', *The Critic*, March 2024. Accessed at: https://thecritic.co.uk/issues/march-2024/hollowed-out-humanities. See also, Aden Barton, 'Five Theses on the Humanities Crisis', *The Harvard Crimson*, 1 December 2022. Accessed at: https://www.thecrimson.com/article/2022/12/1/barton-humanities-crisis.

5 C. S. Lewis, *That Hideous Strength* (The Bodley Head, 1945).

6 These two quotations are commonly conflated. 'I laugh at fortune' is from a letter to Mr [George] Cumberland, 26 August 1799. 'The Goddess Fortune . . . ' is an inscription written by Blake on his illustration to Dante no. 16: *Inferno*, Canto VII.

7 See Leon Kass, *The Beginning of Wisdom: Reading Genesis* (University of Chicago Press, 2006), pp. 49–51; and Charles Foster, *The Selfless Gene* (Hodder, 2009), pp. 132–40. There is an obvious problem with this analysis – namely God's description of the various stages of creation as 'good', and of the completed creation as 'very good'. That problem, significant if one looks to these observations as the foundation of a theodicy, is beyond the scope of this book. I discuss it in detail in *The Selfless Gene* (ibid). Whatever or whoever was behind this ancient insurgency had started an edge-confounding project of homogenization and globalization. The project was continued at the time of the Tower

of Babel, when the ploy (which perhaps seems like a model for global harmony of the kind celebrated by John Lennon in 'Imagine') was to make all humans speak the same language. God doesn't seem to have shared Lennon's perspective. What I've called in the text the 'Mesopotamian Esperanto' (see p. 38) was corrupted by divine fiat, mutual comprehension became impossible, and the presumptuous people were scattered across the Earth. See Genesis 11:9.

8 Per Durrell's friend Robin Fedden, cited by Ian S. Macniven, *Lawrence Durrell: A Biography* (Faber, 1998), p. 314.

9 John Maddox, 'A Book for Burning?', *Nature*, Editorial, 24 September 1981.

10 'John Maddox on Sheldrake and Book Burning', https://www.youtube.com/embed/QcWOz1xjtsY.

11 *The First Australians Fought Back: The Secret Country*, directed by John Pilger, 1985.

12 *Commonwealth Parliamentary Debates*, House of Representatives, 26 September 1901, p. 5233.

13 Exodus 22:18.

14 James 1:12.

15 Brian Levack, *The Oxford Handbook of Witchcraft in Early Modern Europe and Colonial America* (Oxford University Press, 2013); Ronald Hutton, 'Writing the History of Witchcraft: A Personal View', *The Pomegranate: The International Journal of Pagan Studies*, 12: 2, 2010, pp. 239–62; Wolfgang Behringer, *Witches and Witch-Hunts* (Polity, 2004); Lyndal Roper, *Witch Craze* (Yale University Press, 2004).

16 See Charles Foster, 'What's So Great About Consciousness?' *American Journal of Bioethics Neuroscience* 12: 2–3, 2021, pp. 140–2.

17 Steven Mithen, *The Prehistory of the Mind: The Cognitive Origins of Art and Science* (Thames & Hudson, 1999).

Index

Index

Index

Index

Index

About the Author

Charles Foster is a *New York Times* bestselling author whose work has been longlisted for the Baillie Gifford Prize, shortlisted for the Wainwright Prize for nature writing, and won the Ig Nobel Prize for Biology and the 30 Millions d'Amis Prize. He is a fellow of Exeter College, University of Oxford, and has particular passions for Greece, waves, the Upper Palaeolithic, mountains and swifts.